Problem Solving
with Computers

Problem Solving with Computers

PAUL CALTER

Mathematics Department
Vermont Technical College

McGraw-Hill Book Company

New York   St. Louis   San Francisco
Düsseldorf Johannesburg   Kuala Lumpur
London   Mexico   Montreal
New Delhi Panama   Rio de Janeiro
Singapore Sydney   Toronto

Library of Congress Cataloging in Publication Data

Calter, Paul.
  Problem solving with computers

  Bibliography: p.
  1. Electronic data processing—Engineering.
  2. Electronic data processing—Mathematics. I. Title.
  TA345.C35   620'.0028'54   72–1788
  ISBN 0–07–009648–1

PROBLEM SOLVING WITH COMPUTERS

  34567890KPKP798765

To Peggy, Amy,
and Michael

Contents

Preface

The purpose of this book is to bring together in one place the various facts and techniques that a student will need in solving technical problems with the aid of a computer. Many of these things, such as the writing of algorithms, the drawing of flowcharts, the learning of various computer techniques such as writing loops and subroutines, and the various mathematical methods that are particularly applicable to computer use, will be new to those students who have not used a computer before. On the other hand, many of the steps needed to solve a problem with the aid of a computer are the same as those steps needed for solving that problem by any other method. One of these steps, the defining and setting up of a problem, receives particular attention in this book. Once a problem is properly formulated, most students have relatively little difficulty in carrying out the solution, but the setting up of a problem from the original statement is often a major obstacle. Also, the careful checking of results is one area that is usually overlooked, whether the computation was carried out by computer or otherwise. I have therefore devoted an entire chapter to techniques to enable the student to determine the reliability of his answers, to find his errors, and, one hopes, to avoid errors in the first place.

This book is intended primarily for students of engineering or engineering technology who are taking a one- or two-semester course in computer programming. It should also prove useful for those engineers, like myself, who graduated before the widespread availability of computers on the campuses and must now "catch up." The book is not intended for students majoring in computer programming.

In a one-semester course the student should have no difficulty in completing Part 1, including several major projects from Chap. 6. In addition, some selected topics from Part 2 should be included. The remainder of Part 2 can be kept as reference material. The usual course in college algebra is all that is needed to understand anything in Part 1; some knowledge of calculus is necessary to understand most of the material in Part 2. This can be presented at the same time the student is taking calculus.

I have chosen not to tie the text to any particular computer language. Therefore, in order to write actual programs (which I strongly urge), the student will need a user's manual describing in detail the particular language which is available to him.

The practice material included is of two kinds. There are exercises sprinkled throughout the chapters, there is an entire chapter of projects (Chap. 6), and there are projects at the end of every chapter in Part 2. The projects are chosen from many fields of technology and are intended to be difficult enough so that solution by any other means but computer becomes impractical. Many of the projects are intended to be done in consultation with other technical departments. By doing this, it is hoped that the student's course in computer programming and his work in his other technical courses will reinforce each other.

It gives me great pleasure to acknowledge the help I received from many people in the preparation of this text. For their thorough reviews of the manuscript and for many useful suggestions, I am indebted to Dr. Ellis Blade, N.Y.C. Dept. of Air Resources; Walter Granter, Vermont Technical College; Francis Nestor, Wentworth College of Technology; Benjamin G. Klein, New York University; Wayne Scott, Chattanooga State Technical Institute; and Richard Krikorian, Westchester Community College. Mrs. Pauline Gast, Vermont Technical College, gave many useful suggestions for improving the readability of the text; Robert Wonkka, Vermont Technical College, provided the Fortran programs used in the text; and George Fischer, General Electric Company, gave useful criticism of the early chapters. I would like to thank my students who ran and debugged most of the programs in the text, particularly Donald Keller, George Rock, Paul Hodge, Carrol Lawrence, and Kevin Perry. I am also indebted to Miss Julie Wallen for her careful and accurate typing of the manuscript and to my wife Peggy for her typing of the draft and for her constant encouragement.

Paul Calter

Problem Solving
with Computers

PART ONE

Procedures for Problem Solving

1

INTRODUCTION

People have been solving hard problems long before we had computers. There are many techniques and aids for doing a computation, and you should choose the most suitable tool for your problem. A machinist will not use the most precise milling machine in the shop each time he has to cut a piece of metal. He will often find that a hacksaw is better. Some of the tools and aids you might use, aside from working longhand, are the slide rule, logarithms, adding machines, desk calculators (some of which have memories and can be programmed to some extent), graphical techniques, and mechanical devices such as the planimeter. With all these other choices, most of which are simpler, cheaper, and more often at hand, how shall you decide when to use a computer?

1-1 WHEN TO USE A COMPUTER

Use a computer if your problem has one or more of the following features:

1 Great length, which would require a lot of time if done by other methods.
2 Recurring problems, where the same program can be used at a later date, perhaps with slight changes or with different data.

3 Looping, where the same set of operations must be repeated for a number of different cases during the computation.

4 High accuracy needed.

5 Storage of lots of data or instructions for some period of time.

6 Printout of quantities of data which would otherwise require laborious typing and checking but which can be automatically printed in neat tables by the computer.

7 Operations with long lists or large tables of data.

8 Automatic decision making of a simple nature, usually involving the comparison of two numbers as the basis for the decision.

9 Existence of programs especially designed to solve your type of problem, which are already written and corrected. These may be in the library of your computation center or may come with the computer as part of the "software."

The high-speed computer has also made common some techniques that would never be used by hand because of the labor involved. We shall look at *iteration techniques*, where a solution is approached by a series of successive approximations (see Sec. 7-2), and *Monte Carlo methods*, using random numbers (see Sec. 12-3).

You probably know that there are many different types of computers. The methods in this book apply only to the general-purpose digital computer. We shall not be concerned with the various types of analog computers or with those digital computers designed for special tasks, such as navigation or machine-tool control.

1-2 STEPS IN SOLVING A COMPLEX PROBLEM

Not all technical problems are, or should be, solved in exactly the same way, with the same steps always done in the same order. It may be useful, however, to outline in a general way the usual steps followed in the solution of a typical problem. Such an outline will help you understand the layout of this text. It would also be a useful guide in the solution of a particular problem if you are baffled as to how to proceed. This outline is given in the form of a *flowchart* (Fig. 1-1). Flowcharts are described in more detail in Chap. 4, but you do not have to refer to that chapter to understand this simple diagram.

Your first step in the solution is to completely understand the problem. This means understanding clearly what is given and precisely what is required as an answer and having some idea how to proceed from the given to the required. In simple problems you may see all this at a glance, but for more complex cases you may have to "digest" the problem by following the step-by-step procedure given in Chap. 3.

The next few steps are concerned with putting the problem in computable form.

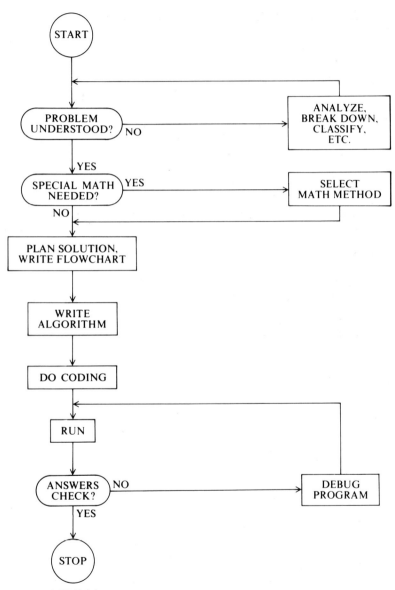

FIGURE 1-1
Steps to follow in solving a complex problem.

The mathematics you need in the solution may range from easy to awful, depending on the problem. In addition, there are many techniques especially suited to computer solution which are not given in the usual college mathematics courses. These are generally referred to as *numerical methods*. A survey of the most useful of these for practical programming is given in Part 2 of this text, along with some typical applications. Become familiar with these methods so that you at least know which techniques are available, even though the exact details are not memorized. Thus you will be able to select a mathematical method suited to your problem and can then refer to this or other texts on numerical methods for the details of execution.

You can now proceed to make a detailed plan of the solution, working out the logic with the aid of a flowchart where necessary, as described in Chap. 4. Then write a step-by-step set of instructions for carrying out the solution. This is called an *algorithm*. If you are familiar with the language of the computer on which this problem is to be run, you can write the algorithm directly in that language. If not, the additional step of *coding* must be performed, perhaps by another person.

Although the writing of the algorithm is much more important and difficult than the task of coding, I strongly urge that you learn a programming language, do your own coding, and run the programs. Only in this way will you become aware of the full power of the computer and learn many of the subtler points of programming.

The problem is now in computable form, and the resulting program can be run. If there are errors in the program which prevent it from running, the program must be debugged and corrected and run again. If the program runs, and answers are printed, there is no guarantee that these numbers are correct. Some sort of checking should always be performed and the program corrected if necessary. The checking of results and the detection of errors are discussed in Chap. 5.

WRITING SIMPLE PROGRAMS

Before looking at complex problems requiring clever logic or special mathematical techniques, let us first write algorithms for some simple, almost trivial computations. The methods we learn can then be extended later for more difficult cases. We shall learn to write a clear step-by-step set of instructions for carrying out a computation. Such a set of instructions is called, as we said in the previous chapter, an *algorithm*. When an algorithm is written in the precise language understood by the particular computer which will be used (Basic, Fortran, Algol, etc.), it is usually called a *program*.

The tasks of defining and clarifying the problem and the writing of a proper algorithm (and possibly the actual program as well) are usually referred to as *computer programming*. The narrower and somewhat mechanical task of rewriting an algorithm in the exact and precise form required by a particular machine is referred to as *coding*.

In this chapter, as elsewhere in the book, the emphasis is on programming, for in order to explain the details of coding we would have to restrict ourselves to a particular computer language. This is not to suggest that the coding of a problem is to be ignored, for you will make little progress unless you actually do your own coding, run your programs on a computer, and analyze your results. This chapter will bring you to the stage where you can write an algorithm for the solution of simple problems. From that, with the aid of a user's manual for your computer, it should not be difficult for you to write the final program. When you become more familiar with that language, you can write the algorithm directly in the computer language. For simple programs it is often possible to omit the writing of a flowchart as well.

2-1 FUNCTIONAL DESCRIPTION OF A COMPUTER

It is not necessary for you to know how a computer works in order to be able to use one, any more than it is necessary to know how an adding machine or a desk calculator performs the various arithmetic operations. If you know as little electronics as I, you should be happy to hear we shall not go into a detailed discussion of computer hardware. It is useful, however, to have a general idea of how the machine works, as it will aid you in writing programs. Let us try to describe the functioning of a computer by means of an inaccurate, incomplete, but rather useful analogy.

Imagine a room equipped with a desk calculator and a typewriter, staffed by a charming young lady who is eager to follow all our (computational) instructions, as in Fig. 2-1. These instructions (the *program*) must be carefully typed in a way that she will clearly understand and are presented at the window in the room marked "input." She will follow the instructions in the program, type the answers as directed in the program, and pass the results out the output window.

An effective computer must have a memory. Suppose that our young lady cannot memorize the mass of information and data necessary for most jobs. Consequently, let us cover one wall of the room with small boxes, similar to those letter boxes found in post offices. This will be the *memory bank*. Each of these boxes, or *memory cells*, has room for only one number at a time, and each cell has a place for a label (the *address*) on the outside. At the start of a run all the labels are removed. Only certain types of labels are permitted; these are legal variable names of the language being used.

It is important to know just how these memory cells are used. Suppose we give the instruction: Let $X = 5$. Upon receiving this instruction, our operator will first scan the cells to see if there is one labeled X. If there is, she will *discard* the number presently in the cell and insert the number 5. Suppose, now, we give the instruction: Print X. She will locate the cell labeled X (there can be only one), look at the number in the cell *without removing or discarding it*, and type the number on the output sheet. Note that when we put numbers into the memory we "write over" and *destroy* any previous number at the same address, but when we use, read, or print a number from memory, we do *not* destroy the number. This is sometimes referred to as *destructive read-in* and *nondestructive read-out*.

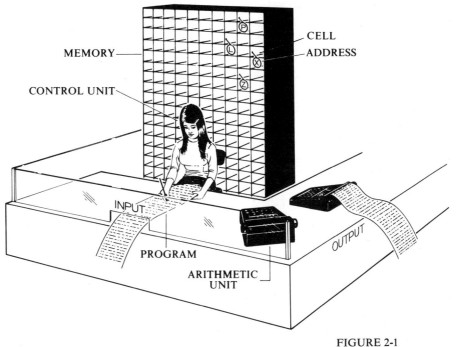

MEMORY

CELL

ADDRESS

CONTROL UNIT

INPUT

PROGRAM

ARITHMETIC
UNIT

OUTPUT

FIGURE 2-1
Diagram of a "computer."

2-2 A SIMPLE PROGRAM

To further illustrate the functioning of the computer, let us see how a simple program would be run.

Let us suppose we wish to find the area of a circle whose radius is 15. Our instructions (algorithm) might look as follows:

Line A   Let $R = 15$
    B   Let $A = 3.1416\ R^2$
    C   Print A
    Z   End

The line letters tell the operator the *order* in which to do each instruction. She will start with the lowest letter and proceed to the highest (which will always be Stop or End) in alphabetical order. We shall use capital letters, omitting O because it looks like a number, and then italic capitals if we run through the alphabet. We shall also omit the word "Line" before each line letter from now on.

What would our charming-lady-computer do with these instructions? She would first find an empty cell, label it R, and insert the number 15. Going to line B, she would then read the number in the cell marked R (15) and square it. She would then multiply the square by 3.1416, using her desk calculator, and place the result in a new box which is then labeled A. The third instruction would ask her to look at the number in the box A and to type this number on the output sheet. Line Z ends the run. Note that at the end of the run we still have numbers in the boxes R and A. Box R still contains 15 and box A contains the product of 3.1416 and R^2.

As an example of what this program would look like when written in an actual computer language, here it is in Basic:

```
10   LET  R = 15
20   LET  A = 3.1416 * R ↑ 2
30   PRINT  A
40   END
```

and in Fortran IV:

```
     R = 15.0
     A = 3.1416 * R * * 2
     PRINT 10, A
10   FORMAT (F 8.4)
     STOP
     END
```

Do not confuse the algorithms with the occasional programs given in this text. You can easily spot the *algorithms* because only they have letters at the start of each line. Also, all *programs* will be printed in capital letters.

EXERCISE 2-1

1 Write a program to compute and print the value of x in the following equations:

(a) $x = \dfrac{3.52(8.47)}{9.23} + 11.49$ *Ans.* $x = 14.7$

(b) $x = \sqrt{297.3}$ *Ans.* $x = 17.24$
(c) $x = (1.472)^{3.721}$ *Ans.* $x = 4.215$
(d) $x = \ln 3.977$ *Ans.* $x = 1.381$

2 Write a program to find the volume and surface area of a sphere whose radius is 3.593 in. Use the equations $A = 4\pi r^2$ and $V = \frac{4}{3}\pi r^3$.   *Ans.* $A = 162.2$ in.^{2}; $V = 194.3$ in.^{3}

3   An object is thrown vertically upward with a velocity V_0 of 50 ft/s.   Write a program to find the height of the object above its launching point after 2.193 s of travel.   Use the formula $S = V_0 t - \frac{1}{2}gt^2$.   *Ans.* $S = 32.22$ ft

4   Write a program to find the radius, surface area, and volume of a sphere if the weight of the sphere and density of the material are given. Write the program and compute these quantities for an iron sphere weighing 11.46 lb.   Take the density of iron as 450 lb/ft³.
                         Ans. $r = 2.190$ in.; $A = 60.28$ in.²; $V = 44.01$ in.³

5   Write a program to convert angular velocities from revolutions per minute to radians per second.   Use it to find the speed of rotation in radians per second of a shaft rotating at 1800 r/min.   *Ans.*   188.5 rad/s

6   Write a program to convert temperatures given in degrees Celsius to degrees Fahrenheit. Convert 68.46°C to Fahrenheit.   *Ans.*   155.2°F

2-3   BRANCH INSTRUCTIONS AND LOOPS

For the simple problem of finding the area of *one* circle, which was done in the preceding section, it hardly pays to use a computer.   A desk calculator would do as well and would be much cheaper.   Suppose, however, we extend the problem a bit.   What if we wanted the areas of the 100 circles whose radii are the integers from 15 to 115?

One way we might write this algorithm would be to write 100 *sets* of statements of the type used in the previous section, with a different radius used in each step.   This would work but we would find it no easier than doing the 100 computations on a desk calculator.   A better way is to write the program so that *one set* of instructions is used 100 times, each time with the proper radius.   Such an arrangement is called a *loop*.

An instruction that causes the computer to leave the written order of instructions and to "jump" to some other instruction is called a *branch instruction*.   Branch instructions are used to form loops.   Consider the following algorithm:

A   Let $R = 15$
B   Let $A = 3.1416 \, R^2$
C   Print A
D   Let $R = R + 1$
E   Go To Line B
Z   End

To the algorithm of Sec. 2-2 we have added lines D and E.   Instructions of the type given in line D are not to be read as algebraic equations for they would certainly be meaningless as such.   That line really means to take the number in the cell R, increase it by 1, and return the new value to the same cell.

Line E causes a return to line B instead of proceeding to the next higher line, Z.   Line E is thus a branch instruction: It causes a departure from the numerical order of instructions.   Moreover, it is an *unconditional branch* because it *always* causes a return to line B whenever line E is reached.

We have thus made a loop (between lines B and E), and this program will indeed give us the areas of the circles whose radii are the whole numbers from 15 to 115. But do you see that we have not told the computer to stop after the one-hundredth computation? It will continue looping until the machine shuts off (many have a built-in time limit) or until we get suspicious that something is wrong and turn it off. To correct this, let us change the program as follows: Change line E to read

> E   If *R* Is Less Than 116, Go To Line B

With this change the computer will return to line B only if *R* is less than 116.   If *R* is equal to or greater than 116, line E will be ignored and the computer will proceed to line Z which ends the run.   This type of branch is called a *conditional branch* because a loop is formed only if certain conditions exist.   A conditional branch always involves a *decision*.   The usual decision is whether one number is larger than another number.

Some of the ways in which two numbers *a* and *b* can be compared, along with the usual algebraic symbols for such comparisons, are as follows:

a equals *b*	$a = b$
a is greater than *b*	$a > b$
a is less than *b*	$a < b$
a is greater than or equal to *b*	$a \geq b$
a is less than or equal to *b*	$a \leq b$
a is not equal to *b*	$a \neq b$ or $a <> b$

When these comparisons are made in a program, *a* will be some particular number stored in the memory and *b* will be some other number, perhaps also drawn from memory.

Our program now contains the *four essential elements* needed for any loop. They are:

1   Instructions to give the *initial values* of the variables in the problem (line A in our example)
2   The *body* of the loop, consisting of those instructions that are to be reused each time (lines B and C)
3   Instructions to *change the values* of some variables before going through the loop again, for it makes little sense to repeat the computation with the same numbers (line D)
4   Instructions to *exit* the loop (line E)

In the two sections to follow, we will discuss in more detail the changing of the variables between successive computations and additional ways of coming out of loops.

We can now complete our algorithm:

A   Let $R = 15$
B   Let $A = 3.1416\ R^2$
C   Print A
D   Let $R = R + 1$
E   If $R < 116$ Go To Line B
Z   End

You can probably see that these instructions could have been written somewhat differently and still achieve the same result. For example, we could have written the algorithm as follows:

A   Let $R = 15$
B   Let $A = 3.1416\ R^2$
C   Print A
D   If $R = 115$ Go To Line Z
E   Let $R = R + 1$
F   Go To Line B
Z   End

There are many other ways that you can write even this simple program; for complex programs the number of variations is almost endless.

EXERCISE 2-2

1   Write a program for finding the volumes of spheres whose radii are the integers from 10 to 20.   *Check* $V = 11{,}494$ when $r = 14$
2   Write a program for putting a list of numbers in numerical order.
        Hint   Repeatedly go through the list, interchanging adjacent numbers that are not in order. Stop the run when no more interchanges are possible.
3   Write a third variation to the algorithm used as an example in Sec. 2-3.
4   Write a program that will produce a table of cubes of the numbers from 10 to 20 in steps of 0.5.   *Check* $15^3 = 3375$
5   A purchasing agent for a certain manufacturing company always requires three bids on any item to be bought and always buys from the middle bidder. Write a program that will pick out the middle value from a set of three numbers.

6   Write a program for finding $\sqrt{x}$ by taking successive approximations.   Compute $\sqrt{2}$ correct to four decimal places by this method.   *Ans.*   1.4142

> *Hint*   Take a guess at $\sqrt{2}$ and divide 2 by your guess.   If the square of your guess is different from 2 by more than the fourth decimal place, take as a second guess the average of your first guess and the quotient.   Keep repeating this process (iterating) until the correct value is obtained.

7   Write a program to pick out the largest number in a list of numbers and its position in the list.

8   A manufacturer makes a line of tubing ranging in inside diameters of 1 to 5 in., in steps of $\frac{1}{4}$ in.   Write a program to find the cross-section area for each tube size.

Check   Area $= 7.0686$ in.^{2} when $D = 3$

9   Write a program which will have the computer print out a table of points on the curve

$$r^2 = 9 \cos 2\theta \quad \text{(lemniscate)}$$

Use these points to plot the curve by hand on polar coordinate paper.

Check $r = \pm 2.121$ when $\theta = 30°$

10   A certain nuclear reactor of circular cross section has a power output of 200 kW/cm^3 at its center.   As we move radially away from the center, the power output decreases, following a cosine function, to a value of 25 kW/cm^3 at the edge of the reactor, at a radius of 220 cm from the center.   Write a program to compute the power output at radii between 0 and 220 cm in steps of 22 cm and use the computed values to plot the radial power distribution on rectangular graph paper.

Check   Power output $= 150$ kW/cm^3 when $r = 110$ cm

11   The displacement x of a certain piece of machinery under vibration is given by

$$x = e^{-4.12t}(0.004 \sin 16t + 0.002 \cos 16t)$$

where x is in inches and t is the time in seconds.   Write a program that will print a table giving x for values of t from zero to 2 s in steps of 0.1 s.

Check $x = 4.72 \times 10^{-4}$ in. when $t = 0.4$ s

2-4   LOOPS WITH DATA STATEMENTS

It is always necessary to change the value of one or more variables before the computer goes through a loop once more, as the radius of the circle was changed in our previous example.   If the successive values of these variables form a *sequence*, that is, follow a *definite pattern* or law of formation, we may compute these values with statements such as:

Let $B = B + 2$
Let $X = 3X$
Let $W = 5W - 4$

Suppose, however, that our variable is to take on values that cannot be generated? For example, suppose we want to find the areas of the circles having radii of 1.385, 2.395, 3.539, ..., 99.937. This string of numbers follows no orderly pattern and hence is not a sequence. A number in this string cannot be found from the one before it. Since we cannot generate the required radii we must store all these numbers in the memory of the computer and draw them out as they are needed. Referring to our letter-box analogy for the memory bank, we can imagine that a certain group of cells is reserved for data of this sort. Our operator will place each number in a separate cell, keeping them in the same order in which they appear in the program. At this time, however, no names are assigned to these cells. This group of cells will be called the *data block*. When instructed to *read* a number, the operator will take the first number in the group and give it a name, as directed in that same instruction. This number can then be used as needed later on. The next time she is told to read a number from the data block she will assign a name to the second number, as directed in the second READ instruction. If, as in a loop, the second READ instruction is really the first one used a second time, the first number will be thrown out and the same name will be given to the second number. Thus she proceeds through the data block, taking the numbers in the order in which they were originally presented in the program.

You will see how these data are used in a program containing a loop in the following algorithm. It is for finding the area of those circles having radii given in a table of data.

A   Store The Numbers 1.385, 2.395, 3.539, ..., 99.937
B   Read A Number From The Data Block.   Call It R
C   Let $A = 3.1416\ R^2$
D   Print R And A
E   If $R < 99.937$ Go To Line B
Z   End

In line B we assume that the numbers will be read in the order of line A, starting with the first number given.

From now on let us write instructions such as line A simply as:

A   Data 1.385, 2.395, ..., etc.

and line B as

B   Read R

This would be a good time for you to refer to your user's manual to find the precise statements for storing and reading data in the language available to you.

In Basic the above program would appear as follows:

```
1   DATA 1.385, 2.395, 3.539, ..., 99.937
2   READ R
3   LET A = 3.1416 * R ↑ 2
4   PRINT R, A
5   IF R < 99.937 THEN 2
6   END
```

and in Fortran IV,

```
1   READ 2, R
2   FORMAT (F 7.3)
    A = 3.1416 * R * * 2
    PRINT 3, R, A
3   FORMAT (F 7.3, F 10.3)
    IF (R − 99.937) 1, 4, 4
4   STOP
    END
    1.385
    2.395
     ⋮
    99.937
```

EXERCISE 2-3

1 Write a program to find the volumes of the spheres having the following radii: 1.493, 2.729, 3.572, 4.284.   *Ans.*   13.94, 85.13, 190.9, 329.3

2 A manufacturer makes cylindrical storage drums with the following outside dimensions (inches):

	Radius	Length	Wall thickness
Drum A	11.32	19.36	0.09
Drum B	18.66	27.49	0.16
Drum C	22.38	35.84	0.21

Write a program to find the inside volume (in gallons) of each of these drums.

Ans.   32.89, 126.4, 236.7

3 Write a program that will find the prime factors of any integer.

Check   The factors of 9990 are 2, 3, 3, 3, 5, and 37.

4 Write a program to find and print all the prime numbers between 0 and 100.

2-5   WAYS TO EXIT A LOOP

In the example of Sec. 2-3 we ended the looping by the instruction

E   If $R < 116$ Go To Line B

There, the magnitude of our independent variable was examined to see if it was *outside a given range.*   In the example of Sec. 2-4 we ended looping with the instruction

E   If $R < 99.937$ Go To Line B

Here again the computer left the loop when our independent variable got out of range.   In this section you will see a few more ways to end a loop.

Suppose that, for the example of Sec. 2-4, we did not know at the time we wrote the program the values of radii to be used.   You may sometimes have to write a program before the exact data are known and have to add data afterward.   How, then, can you write an instruction to end a loop after the last number in the data statement has been used if you do not know what that number will be?   You may do this by means of a *dummy* number.   Take any number which you are sure will never appear in the actual set of data, and put it at the end of the data block.   In your instruction ending the loop you instruct the computer to watch for the dummy number; when it is reached, the loop will be ended.   Let us rewrite the algorithm of Sec. 2-4, using a dummy number.   We know that negative numbers will never appear in a list of radii, so let us choose, say, -10 as the dummy (a dummy number is also called a *sentinel* or a *flag*).   Our algorithm is then:

A   Data 1.385, 2.395, ..., 99.937, -10
B   Read R
C   If $R = -10$ Go To Line Z
D   Let $A = 3.1416R^2$
E   Print R And A
F   Go To Line B
Z   End

Instead of looking at the independent variable to see if it is still within range, as R was watched in Sec. 2-3, it is sometimes more logical to watch the dependent variable, which was A in that problem.   Suppose, for example, that we wanted to compute the areas of all circles having integers for radii and having areas less than 500 in.^{2}.   Our algorithm might appear as follows:

A   Let $R = 1$
B   Let $A = 3.1416 R^2$
C   Print R And A

D   Let $R = R + 1$
E   If $A < 500$ Go To Line B
Z   End

Notice that here the area, rather than the radius, is watched to see if it gets out of range, as was done in previous examples.

Sometimes we know how many times we want the computer to go through a loop and may use this fact to terminate it.   For example, suppose we wish to compute and print the first 100 terms of the Fibonacci sequence

$$1, 2, 3, 5, 8, 13, \ldots$$

where each number is the sum of the two preceding numbers.   Consider the algorithm:

A   Let $N = 1$
B   Let $F = 0$
C   Let $S = 1$
D   Let $T = F + S$
E   Print T
F   Let $F = S$
G   Let $S = T$
H   Let $N = N + 1$
I   If $N < 101$ Go To Line D
Z   End

In this algorithm we used the symbol N to keep track of the number of times we used the loop.   In line A we initialized N to 1, in line H we increased N by 1 each time we used the loop, and in line I we examined N to make sure that we stopped the looping after the one-hundredth time.   Because of its function, N is called a *counter*.

2-6 NESTED LOOPS

In our previous problem in which we computed the areas of a number of circles, we had only one quantity that we wished to vary (the radius).   We formed a single loop and changed the radius each time the computer went through the loop.

What if we have to vary more than one quantity?   For example, suppose we need the volumes of 1-ft-long cylinders whose radii are the integers from 10 to 20 in., and then had to repeat the computation for 1.5-ft-long cylinders, then 2-ft cylinders, and so on.   We can do this problem by means of *nested loops*.   In one loop we can vary the radius and in another loop vary the length.   Consider the algorithm:

A   Let $H = 1$
B   Let $R = 10$
C   Let $V = \pi R^2 H$
D   Print H, R, And V
E   Let $R = R + 1$
F   If $R \leq 20$ Go To Line C
G   Let $H = H + 0.5$
H   If $H \leq 10$ Go To Line B
Z   End

Here we vary the radius in the inner loop and the cylinder length in the outer loop.   The two loops are bracketed on the algorithm.   Notice that one loop is nested inside the other; one loop must always lie completely inside the other with no overlaps.   Also, each loop must contain all the four elements needed for a single loop that were listed in Sec. 2-3.

EXERCISE 2-4

1   Write a program to find the moment of inertia of beams of rectangular cross section, having depths of 4 to 20 in., in steps of 2 in.   For each depth the beam width is to vary from 2 to 8 in., in steps of 1 in.

The moment of inertia of a rectangular area with respect to an axis parallel to the base and passing through the centroid is

$$I = \frac{bh^3}{12}$$

where b is the width and h is the depth.   *Check* $I = 500$ when $b = 6$ and $h = 10$

2   The velocity of a fluid flowing through a pipe, in feet per second, is given by

$$V = 0.143C(d)^{0.64}$$

where C is the roughness coefficient and d is the pipe diameter, in inches.   Have the computer print a three-column table, giving the velocity for pipe diameters from 1 to 5 in., in steps of 1 in., and for roughness coefficients of 82, 105, and 136.

Check $V = 30.33$ when $d = 3$ and $C = 105$

2-7   BUILT-IN LOOP INSTRUCTIONS

Because loops are needed so often it is quite likely that your machine has built-in instructions to make the writing of loops easier.   Fortran, for example, uses the DO instruction, and Basic uses the FOR and NEXT statements.   You are urged at this point to look at your user's manual and become perfectly familiar with any special loop instructions there. To illustrate the advantage of these special instructions, the example of Sec. 2-6 is now done in Basic, using the FOR and NEXT statements.

```
10   FOR H = 1 TO 10 STEP 0.5
20   FOR R = 10 TO 20
30   PRINT H, R, 3.1416*R↑2*H
40   NEXT R
50   NEXT H
60   END
```

The same program in Fortran IV appears as follows:

```
     DO 30 H = 1, 10, 0.5
     DO 30 R = 10, 20
     A = 3.1416 * R**2 * H
     PRINT 20, H, R, A
20   FORMAT (F 5.1, F 4.0, F 8.1)
30   CONTINUE
     STOP
     END
```

2-8   PARTIAL SUMS AND PARTIAL PRODUCTS

If we wish to add several quantities, all of which are stored in the computer memory, we may do so with a simple statement such as:

Let $S = A + B + C + D +$ Etc.

Suppose, however, that the quantities A, B, C, etc., are not all stored at the same time. They may have been computed one at a time, printed, and then *discarded*. For example, suppose we wanted the sum of the areas of all the circles computed in the example in Sec. 2-3. The area is computed in line B of the program and printed in line C. The radius is then changed and a new area is computed, *destroying the previous value* (remember destructive read-in?).

We may still add the areas by forming a *partial sum* somewhere *within the loop* and *accumulate* the areas as they are computed. When the computer has gone through the loop for the last time, the partial sum will then be the *complete* sum. Our summing instruction (still referring to the example of Sec. 2-3) may be simply:

Let $S = S + A$

The sum S starts out at zero (if the computer does not set all variables to zero at the start of a run, then the statement "Let $S = 0$" must be inserted at the start of the program), and each A is added as it is computed.

Partial products can be formed in a similar way with statements such as

Let $P = PQ$

where Q is the factor by which the product must be increased each time through the loop. Of course, P must be set equal to unity at the start of the program.

EXERCISE 2-5

1 Write a program for computing the square, square root, cube, and cube root for the integers from 1 to 10. Have the results printed in a 5-column 10-row table with appropriate column headings. Also, add the 10 numbers in each of the 5 columns and print these sums below the table.   *Check* Sum of the cubes $= 3025$

2 Write a program to compute the sum of the squares of the even integers from 82 to 106.
          Ans. 115,596

3 A loan of $600 is to be repaid by payments of $30 every three months. Part of the $30 will go toward paying off the loan and part will go toward interest, which will be at an annual rate of 6 percent on the unpaid balance. Have the computer print out a repayment schedule in a five-column format giving:

  1 Payment number
  2 Amount of payment ($30 except for the last payment)
  3 Amount of payment going toward repaying the loan
  4 Amount going toward interest
  5 Balance

The last payment must be adjusted to make the final balance come out exactly zero, with no overpayment. Also compute and print out the total interest paid. Round all figures to the nearest penny.   *Check* Total interest $= \$118.70$

4 Write a program that will find $n!$ for any integer n. [Remember that the symbol $n!$, read "factorial n," denotes the product of the positive integers from 1 to n. Thus, $4! = 4(3)(2)(1) = 24$.]   Check your program by finding 10!.   *Ans.* 3,628,800

5 Compute

$$y = \sum_{x=0}^{n} xn! \qquad \text{for } n = 3, 5, \text{ and } 7 \qquad Ans. \quad 36, 1800, \text{ and } 141,120$$

(See Prob. 4 for a definition of $n!$.)

6 Compute

$$y = \sum_{i=1}^{n} \sum_{j=0}^{i} i(i+j) \qquad \text{for } n = 12 \qquad Ans. \quad 10,101$$

7   Write a program to compute the sum of the first n terms of an arithmetic progression (AP). Use it to find the sum of the first 24 terms of an AP having a first term of 3.493 and a common difference of 5.382.   Do not use the formula for the sum of an AP but have the computer actually add up the terms.   *Ans.*   1,569.26

8   Write a program to compute the sum of the first n terms of a geometric progression (GP).   Use it to find the sum of the first 11 terms of the GP whose first term is 8.45 and whose common ratio is 0.779.   Do not use the formula for the sum of a GP.

Ans.   35.7841

9   Compute the value of e (the base of natural logarithms), correct to five decimal places, by means of the series

$$e = 2 + \frac{1}{2!} + \frac{1}{3!} + \frac{1}{4!} + \frac{1}{5!} + \cdots$$ *Ans.*   2.71828

10   Use the program of Prob. 8 on the geometric progression

$$8, 4, 2, 1, \tfrac{1}{2}, \ldots$$

Compute the sum of this progression, taking as many terms as necessary to convince yourself that the sum of an infinite number of terms is 16.

2-9   SUBSCRIPTED VARIABLES

Up to now, the variable names we have used have stood only for a single quantity. Thus, if we wanted to enter the numbers

$$5, 8, 3, 9, 2$$

into a program, we might say

Let $A = 5$
Let $B = 8$
Let $C = 3$
Let $D = 9$
Let $E = 2$

where each number is represented by a different variable.   If the numbers in the above *list* were all somehow related, being, say, the number of beers drunk by five different students last night, we might choose to represent all these numbers by the single variable B (for beer).   Thus

$B_1 = $ *number of beers drunk by first student* $= 5$
$B_2 = $ *number of beers drunk by second student* $= 8$
$\cdots\cdots\cdots\cdots\cdots\cdots\cdots\cdots\cdots\cdots\cdots\cdots\cdots\cdots\cdots\cdots$
$B_5 = $ *number of beers drunk by fifth student* $= 2$

Note that there is still one symbol to represent each number in the list, but by making them all B's we have shown that they are all related. We may also represent the *entire* list by the *single* symbol B_S, where S tells the position in the list. In the above list, S can be any integer from 1 to 5, and tells us which student we are talking about. Realize that in a computation involving the symbol B_S, the *subscript* S must always be specified. Thus, although B_S represents the entire list, we will be using only one *element* in the list at a particular instant.

The main advantage of this subscripted notation is ease of manipulation; we can usually generate the subscripts automatically and save ourselves the work of writing each one. For example, if we wanted to store the above list in the computer memory we could do it with the algorithm:

A   Let $S = 1$
B   Read B_S
C   If $S = 5$ Go To Line G
D   Let $S = S + 1$
E   Go To Line B
F   Data 5, 8, 3, 9, 2
G   (continuation of computation)

Suppose that the beer party were to continue for two more nights and that someone was sober enough to record the amount drunk by each student. The record might look something like this:

		Night		
		1	2	3
	1	5	3	4
	2	3	1	3
Student	3	8	7	5
	4	9	3	6
	5	2	4	5

We now have a *table* or *array* containing 15 *elements*, arranged into 5 *rows* and 3 *columns*. As with the list, we can represent the entire table by the symbol $B_{S,N}$, where S identifies the student and N tells the night. Thus, $B_{4,3}$ is the number of beers drunk by the fourth student on the third night, or six bottles. To read a table into the computer using such a *doubly subscripted variable* as $B_{S,N}$ requires nested loops, as in the following algorithm:

A   Let $S = 1$
B   Let $N = 1$
C   Read $B_{S,N}$
D   If $N = 3$ Go To Line G

E   Let $N = N + 1$
F   Go To Line C
G   If $S = 5$ Go To Line K
H   Let $S = S + 1$
I   Go To Line C
J   Data 5, 3, 4, 8, 1, 3, 3, 7, 5, 9, 3, 6, 2, 4, 5
K   (continuation of computation)

As an example of a complete computation using subscripted variables, suppose that our carefree students bought beer costing 31 cents per bottle on the first night, 25 cents per bottle on the second night, and, getting low on cash, had to settle for 18-cent beer on the third night.   Let us write an algorithm to find the amount spent by each student for the entire three-night blast.

We can start by reading-in the table which tells how much each student drank on each night.   Let us do it exactly as we did before, with lines A through J of the previous algorithm.   To these we can add lines to read-in the list of beer prices. Thus:

K   Let $N = 1$
L   Read C_N
M   If $N = 3$ Go To Line R
N   Let $N = N + 1$
P   Go To Line L
Q   Data 0.31, 0.25, 0.18
R   (continuation of computation)

where C_N represents the cost per bottle on night N.   Now, the total cost per student is found by multiplying the cost per bottle by the number of bottles drunk on each night, and adding the three nightly costs together.   The cost to student 1 for the first night is then

$$B_{1,1}C_1$$

or

$$B_{S,1}C_1 \qquad \text{where } S = 1$$

or

$$B_{S,N}C_N \qquad \text{where } S = 1 \text{ and } N = 1$$

and for the second night

$$B_{1,2}C_2$$

or

$$B_{S,N}C_N \qquad \text{where } S = 1 \text{ and } N = 2$$

Similarly, for the third night his cost is

$$B_{S,N} C_N \qquad \text{where } S = 1 \text{ and } N = 3$$

Obviously, we can compute the cost to student 1, T_1, by computing $B_{S,N} C_N$ with S equal to 1 and with N taking the values 1, 2, and 3, and then summing the three values. To find the cost to the second student all we have to do is repeat the same computation after setting S equal to 2. Finishing our algorithm, then:

R   Let $S = 1$
S   Let $N = 1$
T   Let $T_S = 0$
U   Let $T_S = T_S + B_{S,N} C_N$
V   If $N = 3$ Go To Line Y
W   Let $N = N + 1$
X   Go To Line U
Y   Print T_S
Z   If $S = 5$ Go To Line Z
A   Let $S = S + 1$
B   Go To Line T
Z   End

EXERCISE 2-6

1 Finish the computation in the example of Sec. 2-9 to find the total cost to each student.

Check   Student 3 spent $3.58

2 A certain distributor handles five items which are manufactured by four different companies at prices shown in the following price schedule:

Mfr.	Price per item				
	Wicket	Sprocket	Frammis	Gimlet	Bodkin
Shoddy Mfg. Co.	$2.35	4.20	6.40	7.15	10.20
Cheapway Inc.	2.20	5.80	6.25	7.30	10.40
Junky Products	2.45	5.45	5.90	7.00	10.15
Crookco Inc.	2.25	4.85	6.15	7.45	10.25

The distributor orders these five items from the four manufacturers in the following quantities:

	Amounts bought				
	Wicket	Sprocket	Frammis	Gimlet	Bodkin
Shoddy	75	60	48	15	14
Cheapway	81	47	55	18	12
Junky	78	59	52	17	10
Crookco	92	24	36	22	16

Prepare and run a program which will provide the distributor with the following information:

1 Total amount the distributor owes to each of the four manufacturers
2 Total amount distributor will pay for each of the five items
3 Total amount for all items or total amount of the purchase order

Check   Total amount for all items = $3948.90

2-10 SUBROUTINES

We have seen that the use of a loop enabled us to use the same set of instructions many times, thus saving us much work in typing the program.   Sometimes, however, we may wish to use a set of instructions several times but cannot use loops.   For example, suppose we wished to find the value of x in the expression

$$x = \frac{9! + 5!}{7!}$$

We must compute the factorials of three different numbers and use them in a computation, and there is no easy way to do this by using loops.

One way to compute the factorial F of some number N is by means of the instructions

N   Let $F = 1$
P   Let $F = FN$
Q   If $N = 2$ Go To Line T
R   Let $N = N - 1$
S   Go To Line P
T   Return

This small, self-contained *subprogram*, or *subroutine*, can be used at *any place* in the main program simply by transferring control to line N with the branch instruction

Go To Line N

The RETURN instruction will transfer control back to the main program, at the line immediately following the one which called for the subroutine.   Note that the value of N must be specified before the subroutine is called.   Our complete algorithm for computing x would then be:

A   Let $N = 9$
B   Go To Line N
C   Let $A = F$

D   Let $N = 5$
E   Go To Line N
F   Let $B = F$
G   Let $N = 7$
H   Go To Line N
J   Let $C = F$
K   Let $x = (A + B)/C$
L   Print x
M   Go To Line Z

N   Let $F = 1$
P   Let $F = FN$
Q   If $N = 2$ Go To Line T   Subroutine
R   Let $N = N - 1$
S   Go To Line P
T   Return

Z   End

Note the use of line M to jump over the subroutine to avoid using it by mistake.

At this point you should consult your user's manual to learn the precise language rules for writing subroutines in your language.   Basic, for example, calls for a subroutine with statements like

20   GOSUB 110

where the subroutine would start at line 110, and uses the instruction RETURN at the end of the subroutine to send the computer back to the main program.   The instructions in Fortran are similar.

EXERCISE 2-7

1 Write a program for computing factorials.   Use it as a subroutine to evaluate

$$\frac{8! + 4!}{9!}$$

Ans.   0.111177

2 Write a program to compute the square root of a number (see Exercise 2-2, Prob. 6). Use it as a subroutine to evaluate

$$\sqrt{\frac{18.2 - \sqrt{37.4}}{\sqrt{19.8}}}$$

Ans.   1.648

3

DEFINING THE PROBLEM

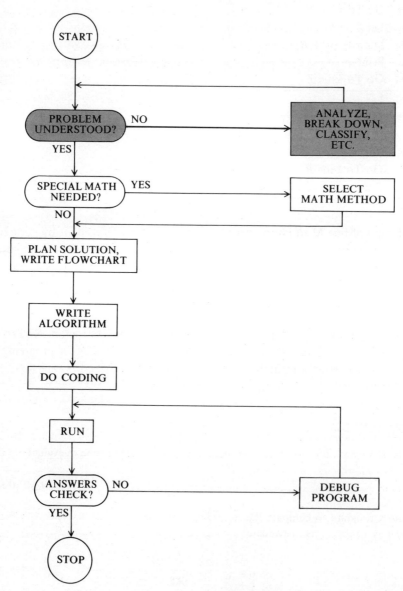

FIGURE 3-1
Steps to follow in solving a complex problem.

When faced with an actual problem which you must solve, you may sometimes see the path to the solution instantly. If you happen to get such a flash of insight, consider yourself lucky and go ahead with the solution. More often, however, a problem will be completely baffling at first glance and you may not even know how to start. Instead of staring at the problem dumbfounded and helpless, there are some simple things you can do to help to clear away the fog in which it seems to be shrouded. When you have completed these preliminary operations, it is hoped you will understand the problem sufficiently to be able to complete the solution. If not, you can resume your staring and wait patiently for an idea.

It should be noted that the steps that we shall outline in this chapter to help clarify a problem are applicable not only when the solution is to be carried out by computer. They are sound steps to follow in approaching any technical problem, regardless of the methods to be used.

3-1 CLASSIFYING THE PROBLEM

Before we can proceed with a solution, it is obvious that we must understand clearly the purpose of the problem. What, exactly, are we trying to find? What is it that we must do? What are we trying to prove? It is often helpful to try to classify the problem, which may fall into one of the following categories.

Solution of one or more equations For example, given the three coefficients in any quadratic equation, find the two roots.

Analysis of an existing design For example, find the load that a given beam can safely carry.

Create a new design For example, determine the dimensions of a beam which can safely carry a specified load.

Optimize a system or a design For example, given a basic lens design, make small changes in the curvatures of the surfaces, spacings, etc., so as to produce a lens having the best possible image quality.

Conversion of a set of data For example, given a table of dimensions in inches, convert to millimeters and print the result.

Operations on lists Some operations often required are searching, merging, ordering, or sorting. For example, given a customer mailing list, remove the names of all customers that have not ordered merchandise in the last year, and print the new list. Or, similarly, given a customer list in alphabetical order, rearrange the list in the order of the total value of purchases during the past year, and print the new list together with the amount spent.

The categories given above do not pretend to cover every situation, and your particular problem may not be represented. In addition, complex problems often do not fall neatly into one category but may overlap several of them. The list will, however, help you to organize your thinking about the problem.

3-2 UNKNOWNS AND DATA

After you know the object of the problem, there are some specific things you can do to further clarify your thinking.

IDENTIFY THE UNKNOWNS What quantities are you seeking? Do you understand exactly how they are defined? List the unknowns. What are the units? Introduce notation, using symbols that can be used unchanged in your computer program and which also suggest the quantities which they represent.

IDENTIFY THE DATA, OR INPUTS What information is given in the statement of the problem? List the data along with units, symbols, and definitions where necessary. Are there some data required for the solution which are not given in the problem statement but which can be found in published tables or handbooks?

3-3 CONNECTING RELATIONSHIPS

Are there any equations, formulas, or physical laws given in the problem statement? If so, list them, using the notation already devised. If they are not given, where might they be obtained? It is here that you must draw upon your knowledge of mathematics, physics, engineering, business, etc. You are hardly expected to memorize every physical relationship, but you should at least know which ones exist in your own field and the conditions for which they are valid. Often the relationship between the data and the unknowns will not be some basic physical law but merely a set of *empirical data* resulting from laboratory measurements of the quantities involved. Later chapters of this book will give techniques for handling this type of relationship with a computer.

3-4 DIAGRAMS AND OUTPUT FORMAT

Whenever possible, make a clear diagram of the situation, labeling all parts.

On a separate sheet labeled Output make a *skeleton tabulation* of the results of the computation, leaving blank those spaces to be occupied by the computed numbers. Not every type of printout will be possible on a particular computer, and the user's manual should be carefully consulted to learn which formats are available and how they are obtained.   The important thing is to be sure that *all* the required information is printed and that all printed quantities are clearly identified.   If your printout capabilities are limited, the information can always be rearranged into an attractive and readable format by means of scissors, paste, and typewriter.

3-5   ASSUMPTIONS

Frequently a problem may be too vague or ill-defined to solve.   There may be "holes" in it, such as missing data, units, or conditions.   This is particularly true on the job, where problems do not come neatly packaged and predigested as in the classroom. When this occurs, the obvious thing to do is to ask questions of whoever gave you the problem, in an effort to clarify any doubtful points.   However, for a number of reasons, you may not get answers, and you should be prepared to make as many assumptions as necessary to enable you to solve the problem.   These assumptions will give the conditions under which your solution is valid.   They should be clearly labeled as assumptions and presented as an integral part of your solution.

Following the steps above will usually bring you to the point where you can see the remaining steps needed to complete the computation.   Note that all these steps may not be needed and that you do not necessarily have to do them in the order given. At this point you should reread the statement of the problem.   Are you providing everything that is required?   With a little extra effort can you provide more than was asked for?

Let us illustrate the use of these steps by means of examples.   The examples that follow were especially chosen to be somewhat vague and to contain terms unfamiliar to most people.

3-6   LENS: AN ILLUSTRATIVE EXAMPLE

Write a program that will choose the correct projection lens and compute the object distance for projecting a 2 × 2 slide onto an 8-ft screen at various projection distances.

Solution

We may classify this problem as one of design, for we are asked to partially design a slide projector.   A bit of a search uncovers the fact that the usual projection scheme consists of a single lens between the slide and the screen, as in Fig. 3-2.

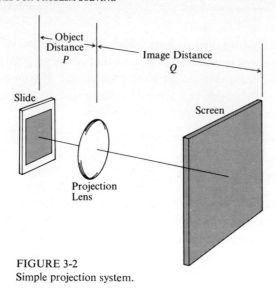

FIGURE 3-2
Simple projection system.

Unknowns We are asked to "choose" a lens. Many factors, such as diameter, quality, price, etc., must be taken into consideration when choosing an actual lens for a given application, but we are not given any information in the problem statement upon which to base such choices. We are told only about the position and size of the image and object, and the lens characteristic that affects these quantities is the focal length. Let us assume, then, that we shall choose a lens only by finding the focal length. Let us give it the symbol F and units of inches.

The *object distance* is an optical term for the distance between the lens and the object which, in our case, is the slide. We choose the symbol P and units of inches.

Data We are told that it is a 2×2 slide to be projected. Only someone familiar with slides would know that these are the nominal outside dimensions of the holder, in inches, and that the actual dimensions of the translucent portion are 0.94×1.36 in.

Since only one screen dimension is given, let us assume that the screen is a square, 8×8 ft.

We are asked to choose the right lens for "various projection distances." Let us assume the projection distance to be the distance from the lens to the screen, which is also called the *image* distance. We choose the symbol Q and units in inches. We are free to specify Q, and so let us take its minimum value as 48 in. and its maximum value as 240 in., assuming that all reasonable projection distances would fall somewhere between 4 and 20 ft. Let us further choose to vary Q in 12-in. steps.

Connecting relationships   If we assume that the thickness of our lens is small compared with the other dimensions of the system, we find that the thin-lens equation

$$\frac{1}{F} = \frac{1}{P} + \frac{1}{Q}$$

or

$$F = \frac{PQ}{P+Q} \qquad (3\text{-}1)$$

provides us with a relationship between the unknown quantities F and P and the known quantity Q.   We have one equation and two unknowns, and therefore we require another equation.   Noting that we have not yet made use of the slide and screen dimensions, we look for a relationship between these quantities and our unknowns and find

$$Magnification = \frac{image\ size}{object\ size} = \frac{image\ distance}{object\ distance}$$

or

$$\frac{8 \times 12}{1.36} = \frac{Q}{P}$$

and

$$P = \frac{1.36Q}{96} \qquad (3\text{-}2)$$

if we assume that the longer dimension of the image is to just fill the 8-ft screen.

We may thus choose Q, compute P from Eq. (3-2), and then find F from Eq. (3-1).

Output format   A three-column table will adequately present all the required information.

Lens-to-screen distance, in.	Focal length of required lens, in.	Slide-to-lens distance, in.
48		
60		
72		
. . .	. . . .	. . . .
240		

Thus our once vague problem has been broken down and clarified to the point where the writing of a program for its solution becomes a simple matter. You should complete the solution, writing a program with a single loop as described in Sec. 2-3. You should find that for a projection distance of 10 ft a lens having a focal length of 1.68 in. (rounded to three figures) is needed.

Let us now tackle a harder problem.

3-7 TRAVERSE: AN ILLUSTRATIVE EXAMPLE

A surveyor lays out a closed traverse, recording the lengths of all courses, the deflection angles between courses, and the azimuth of the first side. Write a program to compute the coordinates of all stations, the error of closure, and the accuracy of the survey.

Here we have a problem statement that would be baffling to anyone with no knowledge of surveying. Not only is the situation new to us, but there are unfamiliar terms and missing units.

Solution

We begin by consulting some surveying texts and discover that a *traverse* is a continuous series of lines called *courses* running between a series of points called *stations*. Furthermore, a *closed traverse* is one that closes upon itself, with the first and last stations being the same point. The purpose of a traverse is to determine the locations of all the stations, relative to one of these stations. The location of a station is given by its distances north and east of the reference station. We also find that, because of inevitable measuring errors while laying out the traverse, the coordinates of the last station as computed from the field data are seldom identical to the coordinates of the first station, although these are actually the same point. The computed distance between the first and last stations is called the *error of closure*. The *accuracy* of the survey is defined as the error of closure divided by the sum of the lengths of all the courses. The azimuth of a side is its angular distance from north, measured clockwise (CW). A *deflection angle* is defined as the change in direction of the traverse at a station. It is considered positive if measured to the right and negative if measured to the left.

Assumptions

1   Since no units are given in the problem statement, let us assume the usual surveying units of feet for length and of degrees, minutes, and seconds for angles (DMS).
2   Let us choose the first station as the reference for measuring the coordinates of the remaining stations.

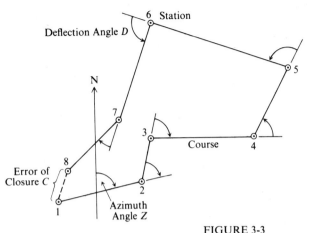

FIGURE 3-3
Definition of terms in a traverse.

Unknowns

1 *Station coordinates*   The distances in feet north and east from the first station.   Choose N and E for symbols.

2 *Error of closure*   The distance in feet from the first station to the computed position of the last station.   Choose the symbol C.

3 *Accuracy*   Dimensionless ratio of the error of closure to the sum of the lengths of the courses.   Choose the symbol A.

Data

1 *Course length*   The distance in feet between two consecutive stations. Choose the symbol L.

2 *Azimuth of first course*   The angular distance CW from north in DMS. Choose the symbol Z.

3 *Deflection angles*   The angle, in DMS, between the prolongation of the back direction and the forward direction; positive if measured to the right (CW).   Choose the symbol D.

Diagram   See Fig. 3-3.

Output format   A skeleton output sheet might look as follows:

Station	Coordinates	
	North	East
1		
2		
3		
4		
etc.		

The error of closure is_____feet.
The accuracy is 1:_____.

Relationships   We are now faced with the more difficult task of finding relationships between the data and the unknowns, for none are given in the problem statement.   We shall obtain these relationships from trigonometry.

Referring to Fig. 3-4 we see that station B is north of station A by an amount

$$Y = L \cos Z$$

and is east of station A by an amount

$$X = L \sin Z$$

If we now add the distances Y and X to the coordinates of A, we will obtain the co-ordinates of station B.   Thus if we maintain partial sums of Y and of X as we go around the loop, these partial sums will be the station coordinates: the distances north and east from the first station.

To compute X and Y we must know the azimuth of each course.   The azimuth of the first course is B.   We can obtain the azimuth of the second course by adding to B the deflection angle D between the first and second courses.   Similarly, the

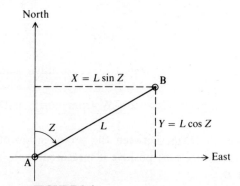

FIGURE 3-4
Trigonometric relationships for the
traverse example.

azimuth of any course can be obtained by adding the deflection angle to the azimuth of the preceding course.

The error of closure is easily computed after the coordinates of the last station are known.

$$C = \sqrt{N^2 + E^2}$$

and the accuracy will be

$$A = \frac{T}{C}$$

where T is the sum of the lengths of all courses.

Our problem has thus been broken down and clarified, with all vague areas now defined. Let us put off the completion of this problem until the next chapter, after we have learned to draw flowcharts, as this will make the job easier.

4

PLANNING THE SOLUTION

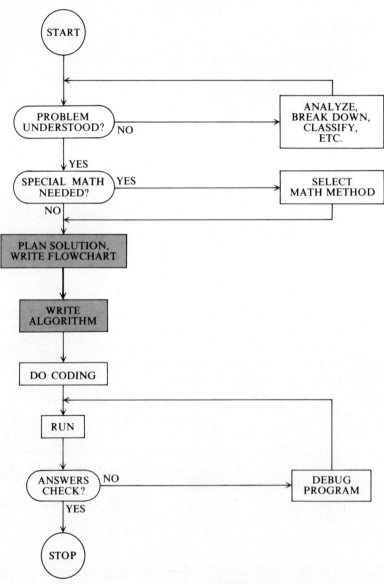

FIGURE 4-1
Steps to follow in solving a complex problem.

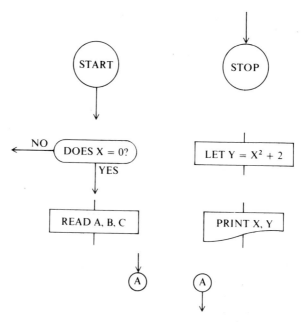

FIGURE 4-2
Flowchart symbols.

4-1   ALGORITHMS AND FLOWCHARTS

Now that we have defined our problem, identified the unknowns and the data, and searched out relationships between them, we can proceed to outline a definite plan for carrying out the computation.   This planning must result in an algorithm: a step-by-step procedure, or recipe, for carrying out the computation.   The best way to write the more difficult algorithms is to work them out first in graphical form. Such a diagram of an algorithm is called a *flowchart*.

A flowchart is a graphical representation of the steps to be followed in solving a problem.   It can show the shortest path through a computation and help you to avoid errors due to faulty logic.   It provides a useful record and can be used to refresh your memory at a later date.   Flowcharts provide a common language among the people working on a large problem and help to coordinate such jobs.

A flowchart is, basically, a string of boxes connected by lines.   Messages in each box tell what is done there; arrowheads on each line give the direction of travel through the chart.   The shape of the boxes is not really important, and any reasonable shapes can be agreed upon by the people working on a job.   This text will use circles for start and stop and also to transfer the flow to a different location without the use of a joining line, rectangles for operations, and decision boxes with rounded sides, as shown in Fig. 4-2.

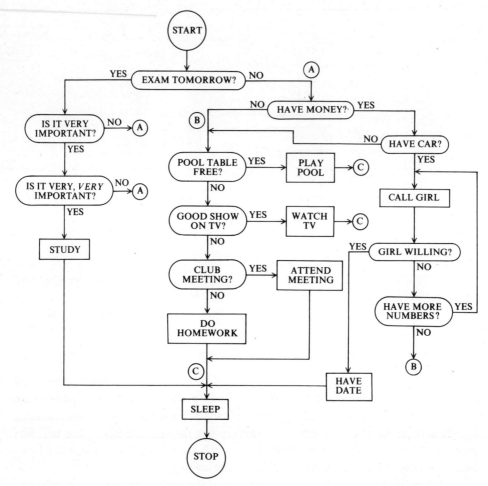

FIGURE 4-3
How to plan an evening.

You have seen one flowchart already (Fig. 1-1) showing the steps to follow in solving a complex problem.   Flowcharts are very useful for planning solutions where decisions must be made, whether in a computation or in everyday situations.   For example, the flowchart in Fig. 4-3 is said to be the work of a student who had trouble planning his evenings.   Strangely enough, his skill at flowcharting did not prevent his flunking computer programming.

Let us now draw flowcharts and write algorithms for a set of problems of increasing complexity.   In each case we shall start with a problem statement.   We then proceed to break down the problem by following the steps given in the previous chapter; then draw a flowchart and finally write the algorithm. The steps we shall follow are summarized here:

Identify, define, label, and list the *unknowns*.
Identify, define, label, and list the *data*.
Find *relationships* between data and unknowns.
Draw *diagrams*.
Choose an *output format*.
Make any necessary *assumptions*.
Draw a *flowchart*.
Write the *algorithm*.

We shall add to this list the final step:

Check the results.

Techniques for the final step are discussed in Chap. 5.

The examples that follow are of a realistic, practical nature and are the type that may be encountered in technical work. You should read the problem statement and try to work the problem by yourself as far as you can, before looking at the solution.

4-2 RESULTANT: AN EXAMPLE WITH NO LOOPS

Given the magnitude (pounds) and direction (decimal degrees measured CCW from some reference axis) of any two coplanar forces, write a program that will compute the resultant of these forces.

Solution

Unknowns The magnitude F (pounds) and the direction A (decimal degrees CCW from the reference axis) of the resultant of the two forces.

Data The magnitudes F_1 and F_2 of the two forces and their directions A_1 and A_2 where the units are the same as for the resultant.

Diagram The two given forces and their resultant are shown in Fig. 4-4.

Relationships One way to find resultants is to resolve each given force into x and y components. The sum of the x components will then be the x component of the resultant and the sum of the y components will be the y component of the resultant.

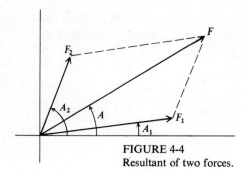

FIGURE 4-4
Resultant of two forces.

The x component of force 1 is

$$X_1 = F_1 \cos A_1$$

and the x component of force 2 is

$$X_2 = F_2 \cos A_2$$

and so the x component of the resultant will be

$$X = F_1 \cos A_1 + F_2 \cos A_2$$

Similarly, for the y component,

$$Y = F_1 \sin A_1 + F_2 \sin A_2$$

The resultant will be

$$F = \sqrt{X^2 + Y^2}$$

and the direction will be

$$A = \arctan \frac{Y}{X}$$

Output format   The output can be simply the two quantities F and A.

Flowchart   The flowchart for the computation is given in Fig. 4-5.   Note the simplicity of the flowchart, containing only a single path which is traveled just one time. It contains no decisions and no loops.

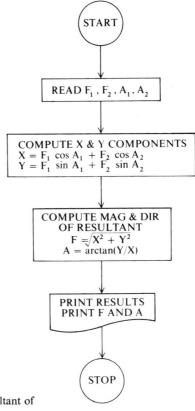

FIGURE 4-5
Flowchart for finding the resultant of
two forces.

Algorithm   With the flowchart to guide us, the algorithm is easily written.

A   Data (insert the four numerical values)
B   Read F_1, F_2, A_1, A_2
C   Let $X = F_1 \cos A_1 + F_2 \cos A_2$
D   Let $Y = F_1 \sin A_1 + F_2 \sin A_2$
E   Let $F = \sqrt{X^2 + Y^2}$
F   Let $A = \arctan (Y/X)$
G   Print F, A
Z   End

This is a good time to consult the section of your user's manual that discusses built-in functions.   The above algorithm requires the sine, cosine, and arctangent functions.   Are these available on your machine?   If so, must the angles be given in radians or in degrees? Will the arc functions give radians or degrees? Will your computer properly handle angles larger than 90°, giving the proper sign to the trigonometric functions, or must this be handled when writing the program?

If the computer you are using interprets all angles as being in radians, rather than degrees, and prints angles in radians, you will have to add instructions to perform the conversions if you wish to work in degrees. To convert degrees to radians, simply multiply by $\pi/180$, and to convert radians to degrees, multiply by $180/\pi$. The problem is somewhat more complicated if the angle to be converted is given in degrees, minutes, and seconds. This conversion may be accomplished with the statement

$$R = \frac{\pi(D + M/60 + S/3600)}{180}$$

where D, M, and S are the number of degrees, minutes, and seconds, respectively, in the angle, and R is the angle in radians. We shall have to perform this conversion in the next example.

4-3   TRAVERSE: AN EXAMPLE WITH ONE LOOP

Let us now complete the problem started in Sec. 3-7. We are at the stage where we may draw a flowchart. First, we must decide how we will get our data into the machine. A data statement is the obvious choice, and the order in which we need the data is, first, the azimuth of the first course; next, the length of the first course; then the deflection angle between the first and second courses; then the length of the second course; and so on. Let us also include dummy values at the end of the list, since we do not know in advance what the last number will be or how many courses the traverse will have. Let us use zeros for dummies since we know that P will never be zero. Now, our angles are all in degrees, minutes, and seconds, and let us assume that our computer works only in radians. To facilitate conversion, let us enter all angles as three separate numbers, P, Q, and R, which will represent the degrees, minutes, and seconds in the angle. Thus for each course we shall type in four numbers.

The flowchart for this problem is given in Fig. 4-6. After P, Q, and R are converted to radians, the result to be called Z, we may then go on to the computation of the displacements X and Y from the previous station. The partial sums E and N will give the coordinates of our present station, and the counter S will keep track of the station number. The partial sum T accumulates the lengths of the sides which will be needed later for the computation of the accuracy of the survey.

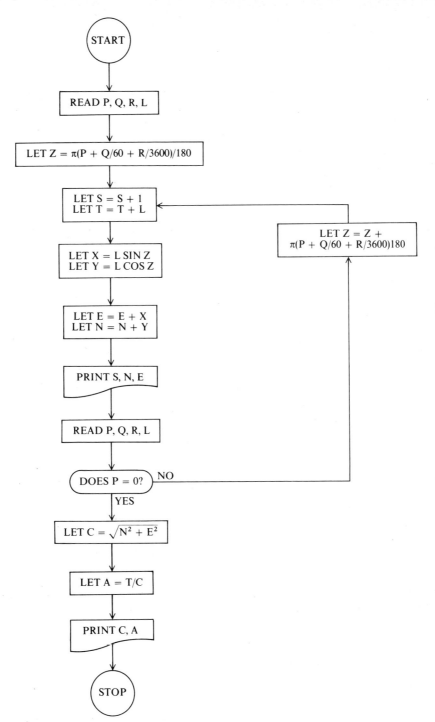

FIGURE 4-6
Flowchart for the traverse problem.

After the station number and coordinates of the present station are printed, we read the next angle and course length. This is first examined to see if it is the dummy, and if it is not, we convert the angle to radians and add it to the azimuth of the previous course. With this new azimuth and new course length we go through the loop again, computing and printing the coordinates of the second station. We will continue the looping until the dummy data are reached, at which time we will leave the loop to compute and print the error of closure C and the accuracy A, thus ending the run. We may now write the algorithm.

A   Read P, Q, R, And L

B   Let $Z = \pi \dfrac{P + Q/60 + R/3600}{180}$

C   Let $X = L \sin Z$

D   Let $Y = L \cos Z$

E   Let $E = E + X$

F   Let $N = N + Y$

G   Let $T = T + L$

H   Let $S = S + 1$

I   Print S, N, And E

J   Read P, Q, R, And L

K   If $P = 0$ Go To Line N

L   Let $Z = Z + \pi \dfrac{P + Q/60 + R/3600}{180}$

M   Go To Line C

N   Let $C = \sqrt{N^2 + E^2}$

P   Let $A = T/C$

Q   Print C And A

Z   End

4-4 CRITICAL: AN EXAMPLE WITH NESTED LOOPS

A gear and a pulley are mounted on a rotating shaft as shown in Fig. 4-7. The gear may assume any one of three positions in order to mesh with other gears in the machine, and, for each of these three gear positions, the pulley may assume any one of three positions in order to be belted to the proper mating pulley. The possible gear locations are 20, 30, and 40 in. from the center of the left bearing, and the possible pulley positions are 60, 70, and 80 in. from the left bearing. Find the lowest critical speed for each of the nine gear-pulley combinations.

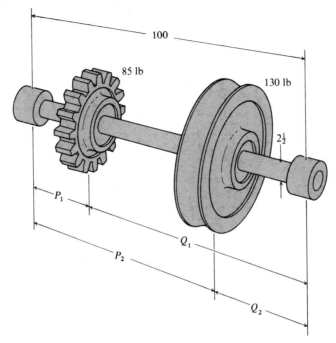

FIGURE 4-7
Gear and pulley arrangement for the critical-speed example.

Solution

Consulting books on vibration theory, we discover that a critical speed is a speed of rotation at which a rotating shaft becomes unstable and large vibrations are encountered. It is important to know them so that they may be avoided. Although trouble may arise at more than one such speed, we are asked to find the lowest of them, or the *fundamental* critical speed.

Unknowns   We seek the critical speed N for each of the nine cases. The most useful units would be revolutions per minute.

Data   Shaft dimensions given are length L, of 100 in., and diameter D, of $2\frac{1}{2}$ in. Gear weight is W_1 (85 lb) and pulley weight W_2 (130 lb). Also, gear positions P_1 and pulley positions P_2 are in inches.

Relationships   From our reference we obtain the equation

$$N = \frac{60}{2\pi} \sqrt{\frac{g(W_1 Y_1 + W_2 Y_2)}{W_1(Y_1)^2 + W_2(Y_2)^2}}$$

where g is the gravitational constant (386 in./s²) and Y_1 and Y_2 are the static deflections of the gear and pulley, respectively.   The static deflections are the amounts that the shaft would sag because of the two weights W_1 and W_2, if it were not rotating (Fig. 4-8).   These are not given, and we must find a way to compute them.

Consulting tables of beam formulas for a beam simply supported at both ends and having a concentrated load at a distance P from the left end, we obtain equations for the deflections anywhere along the beam.   The equations differ, however, for sections to the left of the load, to the right of the load, and directly under the load. In our problem, the load will always be at one of the sections at which we require the deflection.   Furthermore, the equations apply only when there is a *single* load acting on the beam, whereas in our problem we have two loads, and so we shall have to apply the *principle of superposition*.   This principle will allow us to compute Y_1 and Y_2 with only the gear weight bending the shaft, then compute Y_1 and Y_2 with only the pulley weight, and then add the two sets of Y's to get the total deflections with both gear and pulley in place.

Our static deflections, for the two possible cases, are given by

1   Gear acting alone:

$$Y_1 = \frac{2W_1(P_1)^2(Q_1)^2}{6EIL} \qquad Y_2 = \frac{W_1(P_1)(Q_2)}{6EIL} [L^2 - (P_1)^2 - (Q_2)^2]$$

2   Pulley acting alone:

$$Y_1 = \frac{W_2(P_1)(Q_2)}{6EIL} [L^2 - (Q_2)^2 - (P_1)^2] \qquad Y_2 = \frac{2W_2(P_2)^2(Q_2)^2}{6EIL}$$

Output format   The results can be clearly presented in a three-column, nine-row table as follows:

Gear location in.	Pulley location in.	Critical speed, r/min
20	60	
20	70	
20	80	
30	60	
30	70	
30	80	
40	60	
40	70	
40	80	

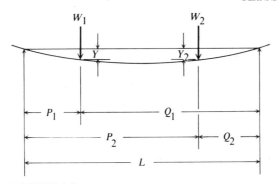

FIGURE 4-8
Static deflections of a shaft.

Assumptions   We are assuming that the bearings act as simple supports with no flexibility, that the shaft weight is negligible, and that the weights of gear and pulley are concentrated.

Flowchart   In the flowchart of Fig. 4-9 we first compute the denominator $D = 6EIL$ as a convenience, as we shall need it later in the equations for Y_1 and Y_2.   Note that the moment of inertia of a circular shaft is

$$I = \frac{\pi d^4}{64}$$

where d is the shaft diameter.

We go on to select the first values of P_1 and P_2 and then compute the distances from the right-hand bearing Q_1 and Q_2.   These values are then used to compute the static deflections Y_1 and Y_2, using equations that are the combinations of the equations for Y_1 and Y_2 given above.   The critical speed N is then computed and printed along with P_1 and P_2 so that there is no confusion as to which case it is applicable. We then examine P_2 to see if it is 80; if it is not we increase it by 10 and go through the computation of N once again.   If P_2 is equal to 80, we know we must now increase P_1 and repeat the computation, after resetting P_2 to its initial value of 60.   Before increasing P_1, however, we first examine it to see if it has already reached its final value of 40; if it has, we may end the computation.

Algorithm

A   Let $D = \dfrac{6\pi(30{,}000{,}000)(100)(2.5)^4}{64}$

B   Let $P_1 = 20$

C   Let $P_2 = 60$

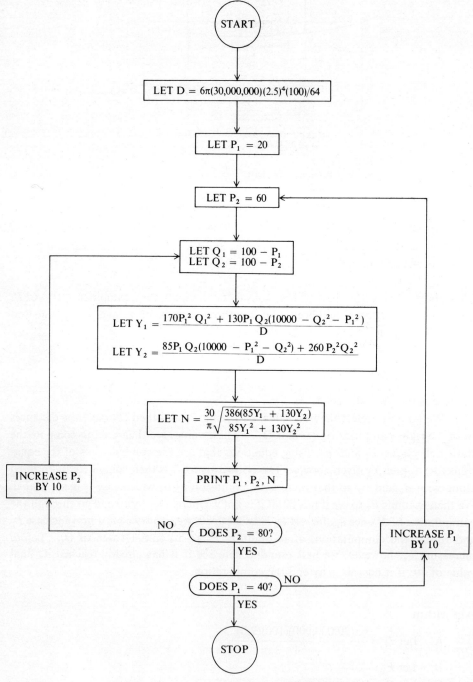

FIGURE 4-9
Flowchart for the critical-speed example.

D   Let $Q_1 = 100 - P_1$

E   Let $Q_2 = 100 - P_2$

F   Let $Y_1 = \dfrac{170P_1{}^2Q_1{}^2 + 130P_1Q_2(10{,}000 - Q_2{}^2 - P_1{}^2)}{D}$

G   Let $Y_2 = \dfrac{85P_1Q_2(10{,}000 - P_1{}^2 - Q_2{}^2) + 260P_2{}^2Q_2{}^2}{D}$

H   Let $N = \dfrac{30}{\pi}\sqrt{\dfrac{386(85Y_1 + 130Y_2)}{85Y_1{}^2 + 130Y_2{}^2}}$

I   Print P_1, P_2, N

J   If $P_2 = 80$ Go To Line M

K   Let $P_2 = P_2 + 10$

L   Go To Line E

M   If $P_1 = 40$ Go To Line Z

N   Let $P_1 = P_1 + 10$

P   Go To Line C

Z   End

5

CHECKING THE RESULTS

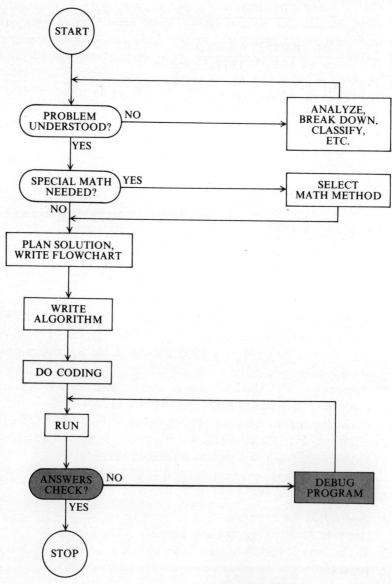

FIGURE 5-1
Steps to follow in solving a complex problem.

At last, after going through all the preliminary steps, you have written the program, entered it into a computer, and run the program. Out comes a set of beautiful answers. Your first impulse will probably be to tear the output sheet from the computer to show it to your boss or teacher and to wait for the praise that will inevitably follow. *Do not do so. Your answers are probably wrong!*

All computations should be checked. If the correctness of your answer is very important, involving perhaps public safety or some other situation where incorrect answers could lead to loss of life, property, your job, etc., then you will check your results by as many different ways as you can devise.

Errors, inaccuracies, and mistakes can creep in at all stages of your work, as shown in the following listing.

5-1   SOURCES OF ERROR

When defining the problem

1 *Poor input data*   Is your source reliable? Were measurement errors made during the collection of these data?

2 *Transcription errors*   Have you copied the data correctly into your program? Have you made the common error of transposing adjacent digits, such as writing 5478 instead of 5748?

3 *Poor choice of mathematical method*   Is your chosen method really right for the job? Does it cause large roundoff or truncation errors? Do such errors add up to such an extent that the answers are worthless? Is the method so unstable that small changes in the input data cause large changes in the results?

4 *Poor assumptions*   Have you oversimplified the problem to such an extent that it no longer resembles the original situation?

5 *Inconsistent units*   Are your units consistent so that they cancel properly? Is the output in the units that are expected? Are you entering angles in degrees into a computer that works only in radians?

When planning the solution

6 *Faulty logic*   Are all the steps in your algorithm in the proper order? Are your decisions and branches correctly set up?

7 *Typing errors*   Have you made more transposition errors while typing the program?

8 *Format errors*   Have you entered an illegal instruction that prevents the program from running?

When running the program

9   *Equipment malfunction*   Has some failure within the computer caused it to produce incorrect results?   Does your teletype or printing unit occasionally "misspell" and type a wrong digit?

10   *Roundoff error*   Do the small roundoff errors generated within the computer when converting to binary accumulate enough to be troublesome?

After the run

11   *Transcription errors*   Have the computer results been properly transferred to other documents?

5-2   CHECKING THE ANSWERS

How do we determine if our answers are correct?   In fact, what do we mean by "correct"?   In most real problems we start with approximate data and use approximate techniques, so that our answers can never be exactly correct.   We must decide just how much accuracy we need.   Fortunately, most technical problems do not require greater accuracy than three or four significant figures, and often less.   The techniques given in this chapter are most suited to finding larger errors, mistakes, and blunders, rather than to refining a computation to achieve high accuracy.   The problem of getting maximum accuracy from a given computation is rather involved and is beyond the scope of this book.   Should the occasion arise where you need accuracy of more than four or five significant figures in a lengthy computation you should consult a text on numerical methods to discover the pitfalls involved.

Let us look now at 11 methods for checking results.   You should use as many of these as necessary to verify the correctness of your work.

1   Rerun the program   An intermittent computer malfunction, or a defect in the printing mechanism, may occasionally cause wrong numbers to be printed, even though the program may be perfect.   To guard against this you can simply rerun the program and compare the two output sheets.   These should, of course, be identical.

2   Estimate the answer   From the nature of the problem you often know roughly how large your answers should be.   Examine your answers.   Do they make sense?   If you are not familiar enough with the problem to know the approximate size of the numbers, devise some way of estimating them.   Make assumptions that will simplify the estimate.   For example, if we had computed the surface area of the complex nose cone of Fig. 5-2, we could approximately check our answer by computing the area of the right-circular cone shown in the same figure by the simple formula

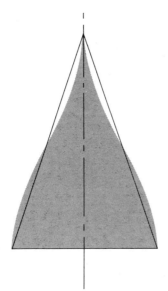

FIGURE 5-2
Aircraft nose cone.

$$A = \tfrac{1}{2}(\textit{perimeter of base}) \times (\textit{slant height})$$

We would not expect the two numbers to agree exactly, but we would not expect a very large difference either.

3   Bracket your answer   If you cannot estimate the quantity for which you are solving, you may be at least able to find two numbers within which your answer must lie.   For example, in computing the length of the bridge cable in Fig. 5-3, you know that the true length must be greater than the distances $AB + BC$, but less than $AD + DE + EC$.

4   Obtain results by a different method   One of the best ways of checking your answers is to completely redo the work by a different method.   If your initial computation was done on a computer, you may be able to write another program which will use a different mathematical technique.   Graphical methods exist for many types of computations; several are mentioned in this book.   They provide an excellent way of checking results.   Mechanical devices such as the planimeter can be used in some instances.

Checking by an alternative method does, of course, involve a great deal of extra work, but in those situations where it is essential that your results be correct, it is well worth the effort.

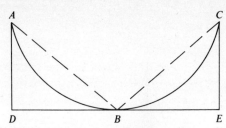

FIGURE 5-3
Suspension-bridge cable.

5 Check by another person It often happens that when a person tries to check his own work he will repeat the same mistake over and over, whereas another person looking at his work will spot the error quickly. For important jobs, have your work checked by someone else, especially when there is no way to check the result by a different method. The "checker" may simply go over your papers looking for mistakes, but it is far better to have him start from scratch and do the entire job, looking at your work only to compare the results.

6 Test cases When you have your program written and running, try it with data for which the results are known or which can be easily found by hand. If your program has branches you should try to select data that will test each branch. As a simple example, if you had written a program to compute the square root of numbers, you would certainly try your program first on numbers for which you know the square root. If you were checking the program which computes the resultant of two forces (Sec. 4-2), you would first try situations for which you know the answers, such as

$$F_1 = F_2 = 1 \qquad A_1 = 0 \qquad A_2 = 90°$$

for which the resultant would be 0.707 at an angle of 45°, or

$$F_1 = F_2 = 1 \qquad A_1 = A_2 = 0$$

for which the resultant would be 2 at an angle of 0°. Many similar test cases could be imagined.

7 Back substitution If your program involves the solution of equations, do not neglect to substitute your roots back into the original equations to see if they check. You can do this manually, or with another small program written just for this purpose.

8 Graphical display It is difficult to spot an error in a large table of numbers. Often, however, when results are plotted, charted, or otherwise visually displayed, the troubles become apparent. Graph your results whenever possible. On the graph you should look for such things as intercepts, asymptotes, maximum and minimum points, points of inflection, general trends, etc. Do the features you observe make sense in the context of your problem? Are there any points that do not fit in with the general trend? Are these points in error, or should the curve actually pass through them?

9 Checks by counting Sometimes errors can be uncovered simply by counting the numbers in your output, in your data inputs, or elsewhere. Are there the correct number of inputs? If you are inserting data in pairs of numbers, is the total number of figures even? Omission of one number causes a mismatch which will make the run worthless. An incorrect number of numbers in the output may indicate faulty logic. These simple and rapid checks should not be overlooked.

10 Check for stability See how sensitive your answers are to changes in the input data by varying the inputs in turn by a small amount. Observe the change in the results. Should large variations occur, you should be suspicious of the answers and conduct further checks. Programs that are too sensitive to small changes in the data are called *unstable* or *ill-conditioned*. Instability may be caused by the nature of the problem, such as trying to find the intersection point of two lines that are nearly parallel. It can also be caused by the way the computation is performed, perhaps with a large buildup of errors.

An estimate of the accuracy of the computation can be obtained by varying each input variable simultaneously in the direction that will cause the answer to increase (or decrease); this amount of change in the answer will be a measure of the accuracy. The input variables should be varied in amounts equal to one-half the least significant figure.

11 Check against prior work You will rarely be faced with a brand new problem, and it is quite possible that a problem similar to yours has already been solved. It may have been done by you at some earlier date or by someone in your company. You may find it in the technical literature or in tables of published data. You should try to compare your results with such outside sources whenever possible. Although you will never find a situation identical to yours, you can at least see if your numbers are consistent with the others. You can usually estimate whether your numbers should be larger or smaller than those with which you are comparing them, and by roughly how much.

5-3 FINDING THE ERRORS

If you should find, as is often the case, that your results do not check or seem suspicious, the next step is to locate and correct any errors in your work. It is quite possible that the steps you perform to check your answers will give you clues to where the mistakes may be, and you should first follow these clues. If the errors are still not located, use the techniques in this section.   They are the techniques of tracing, manual simulation, and the use of unconditional branches.   The removal of errors from a program is often called *debugging*.

PRINT statements are usually inserted into a program to cause the *final* results to be printed.   However, the PRINT statement can be a powerful debugging tool if used at other points in the program to cause the printing of *intermediate* quantities. This technique is called *tracing*, or the *check-point method*.   These intermediate quantities can then be examined, and if they are as expected the PRINT command can be erased and inserted elsewhere.   When the program is working properly, all the intermediate PRINT statements can be erased.

One of the best ways to check your program is to go through it manually and perform all the steps that the computer would perform.   Keep track of all quantities on a sheet of paper.   On this, designate a number of memory cells, giving to them addresses as called for in the program.   The number in each cell is then assigned or changed according to the program.   When a PRINT statement is reached, draw the number from the appropriate cell, and write it on another output sheet.   When "playing computer" in this way, you do not have to work to great accuracy.   Slide-rule accuracy or even estimates are good enough.   This technique is called *manual simulation*, as you are actually acting as the operator in our functional description of a computer in Sec. 2-1.

When checking a long program you can sometimes break it into smaller sections and check each section separately.   You may temporarily eliminate those portions of the program you do not wish to run by using the unconditional branch, or GO TO statement.   This will enable you to "short-circuit" all but the sections you wish to test.   These sections should be provided with any necessary PRINT statements. Make a note of any temporary branch statements inserted, and be sure to erase them when they are no longer needed.

5-4   PRACTICAL TIPS

Listed below are some suggestions to help your prevent making errors, for making your work easier to read and therefore easier to check, and for improving the accuracy of your results.   As with many of the ideas in this book, most of these tips are not limited to computations performed on a computer but apply to other methods as well.

When planning the solution

1   Develop good bookkeeping habits.   Work neatly.   Number and date each page of your computation.   Clearly label each part of your work.

It is a good idea to give each problem a different name and to write this name at the top of each of the sheets that go with the problem, as well as on the program. This is especially valuable when several problems are being worked at once, in order to keep your papers from getting intermixed and for filing material which may be used at a later date. The name, of course, should be chosen so that it suggests the nature of the problem.

Use color for clarity and emphasis. Work in pencil and keep a large eraser handy. Remember that good workmanship is just as important and satisfying when doing a computation as it is in any other worthwhile activity.

2 Carry units along with the numerical values of any physical quantities. Make sure the units cancel properly, making conversions where necessary. Do not supply angles in degrees if your computer works in radians.

3 Carefully check each step as you proceed, instead of waiting until the end.

4 Make accurate and complete diagrams and flowcharts.

5 Choose variable names that will suggest the quantities they stand for so that you will not have to keep referring back to find out what your symbols mean.

6 When writing the algorithm, do not always try to start with the first statement and write the whole algorithm in order; start with the key statements that actually do the computation, and work outward from there.

7 Prepare long programs in sections, or with subroutines, so that these sections may be checked separately.

8 Avoid stringing programs together into one superprogram, for you could easily wind up with a monster that cannot be understood or checked.

9 It is not necessary to turn *all* decision making over to the computer. It is sometimes better to have the machine print out whatever data are needed for the decision and make it yourself. Having too long and too clever a program, with lots of automatic decision making, can result in your "losing contact" with the computation and not really understanding what is going on.

10 Try to write equations so as to reduce the number of operations required. This will improve accuracy as well as reduce the running time. This may be important if the equation is in a loop which is executed many times. Thus the expression

$$ax^3 + bx^2 + cx + d$$

requires eight operations (two exponentiation, three multiplication, and three addition), whereas the equivalent expression

$$[(b + ax)x + c]x + d$$

requires only six (three addition and three multiplication).

11   Beware of situations where you must take the difference of two nearly equal numbers, as this can result in a drastic loss of significant figures. For example, in subtracting 985.3 (four significant figures) from 985.8 (also four significant figures) we get 0.5 (one significant figure). We lost three significant figures in this subtraction. If you cannot avoid such subtractions, try to do them toward the end of the program.

12   When entering data into your program, do not round your better data down to the number of significant figures in your worst data. Enter each number to as many significant figures as it is known. When entering accurately known numbers from published tables, include at least one more significant figure and one more decimal place than your best data. Best accuracy will be obtained if you save all rounding until the end of the computation.

When typing the program

13   Make your program easier to read and to check by inserting remarks or messages at appropriate points.

14   Skip lines to create spaces between sections of a program. Indent certain instructions for clarity. Nested loops become much easier to read if each loop is indented.

When planning the printout

15   If the program is designed to find a single answer to a single problem, it is sufficient to print the one number without any identification. However, most programs result in the printing of many quantities corresponding to different values of the input data. It then becomes important to identify carefully the numbers that come spilling out of the machine so that you know which answers go with which problems, and which quantities they represent, and, often, what their units are. Computers vary in the types of output formats that are possible, and you should study the user's manual for your system to learn what is available. Often a large quantity of data is best presented in a table. If so, appropriate column headings, with units, should be printed first. If input data are varied during the computation, the values of these data used in the particular computation should also be printed. Always be sure to include enough descriptive *words* in your printout so that there can be no chance of wrongly interpreting the results.

16   Do not report answers to a greater precision than the data used to compute them. Round your answers so that they do not contain more significant digits or more decimal places than the least accurate data used in the computation.

EXERCISE 5-1

1 Write a program that will add 0.0001 to itself 10,000 times and will print the sum. How does your result compare with the correct answer of unity? How would you explain any difference?

2 Write a program for computing the sine of an angle from its series expansion

$$\sin a = a - \frac{a^3}{3!} + \frac{a^5}{5!} - \cdots$$

Compute the sine of 30° using only the first five terms of the series, and compute the truncation error (the error introduced by not using all the terms of the expansion). Remember that when using this expansion, as well as the expansion in the following problem, the angle *a* must be expressed in *radians*.

3 Write a program to compute the cosine of an angle from the series

$$\cos a = 1 - \frac{a^2}{2!} + \frac{a^4}{4!} - \cdots$$

Use your program to compute cos 60°, using only the first five terms of the series, and compute the truncation error.

4 Where would you insert PRINT statements in order to achieve a trace of the algorithm in Sec. 4-2?

5 Where would you insert PRINT statements in order to trace the algorithm of Sec. 4-3?

6 Where would you insert PRINT statements in order to trace the algorithm of Sec. 4-4?

7 The following algorithm is for the computation of the sum of the first 10 terms of an arithmetic progression. It is intended that it compute the sums of three separate AP's, having first terms of 4, 9.3, and 115 and common differences of -1, 8, and -5, respectively. Find the errors in this algorithm by manual simulation.

    A   Data 4, -1, 9.3, 8, 115, -5
    B   Read F, D
    C   Let $C = 0$
    D   Let $S = F$
    E   Let $S = S + D$
    F   Let $C = C + 1$
    G   If $C = C + 1$ Go To Line I
    H   Go To Line D
    I   Print S
    J   If $D = -5$ Go To Line Z
    K   Go To Line B
    Z   End

6

PROJECTS

The exercises in this chapter will, in general, require much more time for solution than the exercises given in the previous chapters and are hence called *projects*. They are chosen from many areas of technology. You may have to consult your instructors or other texts in the specific field to clarify the problem and to locate missing definitions and equations. Assumptions will often be required; these should be clearly stated in your solution.

Although data are provided for most of the projects, the computation will be far more meaningful if done with data you have collected yourself in the laboratory. Several of the projects deal specifically with laboratory experiments commonly required in a college physics course.

Many of the projects deal with problems that arise frequently in technical work. If you write your programs so that they can be used at a later date with different data, and clearly label and save them, you will have the beginnings of your personal program library. This could be a great time-saver in the future.

The projects in this chapter do not require any special mathematical or computer techniques beyond what has already been presented. Projects requiring the numerical methods discussed in Part 2 of this text are given at the end of each of the following chapters.

1 Write a program that will take the rectangular coordinates of any point and compute the polar coordinates.   Compute the polar coordinates of the points (5.84,9.02), (79.2,8.47), (−34.4,18.9), (0.288,−1.34), (−6.88,−3.95).

> *Check*   For the first point, 10.7 at an angle of 57.1°

2 Write a program that will take the polar coordinates of any point and compute the rectangular coordinates.   Compute the rectangular coordinates of the following points: 48.2/46.3°; 8.26/147.8°; 0.384/202°; 11.43/314.7°, where all angles are in degrees.

> *Check*   For the first point, (33.3,34.8)

3 Write a program which will compute the resultant of any number of coplanar forces. For each force, enter the magnitude, in pounds, and the direction, in decimal degrees, and have the computer print the resultant in pounds and decimal degrees.

   Use this program to check the results obtained in the physics laboratory with the force table and your graphical and manual computations for the same set of forces.

4 The speed of a projectile can be measured by having it collide with a ballistic pendulum and measuring the distance traveled by the pendulum, either horizontally or vertically. Have the computer produce a calibration chart for a 25-g ballistic pendulum 12 in. long, giving the projectile velocity in increments of 2 ft/s, the pendulum height in inches, and the horizontal distance traveled by the pendulum in inches.   End the tabulation when the pendulum height exceeds 6 in.   The mass of the projectile is 5 g.

> *Check*   Height is 3 in. when velocity is 24 ft/s

5 You are designing an eight-stringed musical instrument to span the octave starting with middle C.   The strings are to be steel (density $= 450$ lb/ft³), 24 in. long, and are to be one of the following commercially available diameters, in inches:

$$0.0320, 0.0359, 0.0403, 0.0453, 0.0508$$

The tension in any string must not exceed 500 lb or be less than 150 lb.   The eight required frequencies (in hertz) are

$$261.6, 293.7, 329.6, 349.2, 392.0, 440.0, 493.9, 523.3$$

The fundamental frequency of a vibrating string is

$$f = \frac{1}{2L}\sqrt{\frac{F}{M}}$$

where L is the string length, F is the tension, and M is the mass per unit length.

   Write a program that will choose the thinnest string meeting the above requirements, and compute the tension in the string, for each of the eight required frequencies.

> *Check*   At 392 Hz, $D = 0.0320$ in. and $F = 192$ lb

6   For the derrick in Fig. 6-1, write a program to compute the tension in the cable and the horizontal and vertical components of the reaction at *C*. Do the computation for different boom positions, letting angle *A* vary from 10 to 80°, in steps of 5°. Assume that the weight of the boom is 2000 lb concentrated at its midspan.

> *Check*   When $A = 30°$, $T = 4510$ lb, reaction at $C = 3750$ lb horizontal and 7495 lb vertical

7   Write a program that will take the coordinates of the end points of the 13 girders in Fig. 6-2, and compute the length of each girder.

> *Check*   The length of girder A = 53.7 ft

8   By means of an acceleration apparatus in the physics laboratory, an object is dropped, and its position is recorded at the end of each of a number of equal time intervals.

   Write a program that will accept these data and prepare a five-column table containing

   a.   Elapsed time
   b.   Position
   c.   Distance fallen in one time interval
   d.   Average speed in one time interval
   e.   Average acceleration in one time interval

9   A steel surveyor's tape must be corrected for variations in length due to changes in temperature. Have the computer print out a correction table, giving the corrections needed at temperatures from $-20°F$ to $120°F$ in steps of $5°F$, and for lengths from 50 to 200 ft in steps of 50 ft. Take the temperature coefficient as 6.45×10^{-6} per degree Fahrenheit, and assume that the tape reads correctly at $68°F$. Round all corrections to the nearest 0.01 ft, and indicate when they are to be added to the tape reading, and when subtracted.

> *Check*   The correction is 0.07 ft at 10°F and 200 ft

10   Write a program to compute the image position, magnification, and image height, if the object position, object height, and focal length are given for a thin lens. Try your program, using a lens having a focal length of 10 cm and with an object height of 5 cm, and letting the object distance vary from 100 to 15 cm in steps of 5 cm.

> *Check*   When the object distance is 60 cm, the image distance is 12 cm, the magnification is 0.2, and the image height is 1.0 cm

Use your program to check values experimentally obtained in the physics laboratory.

11   The steady-state heat flow through a wall made up of several layers in contact is given by the equation

$$q = \frac{\Delta t}{L_a/k_a A_a + L_b/k_b A_b + L_c/k_c A_c + \cdots + L_n/k_n A_n}$$

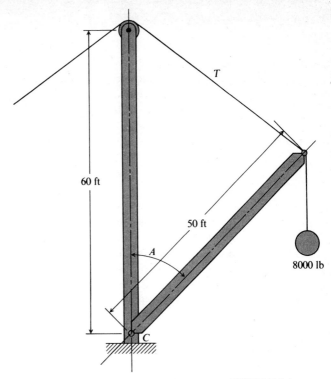

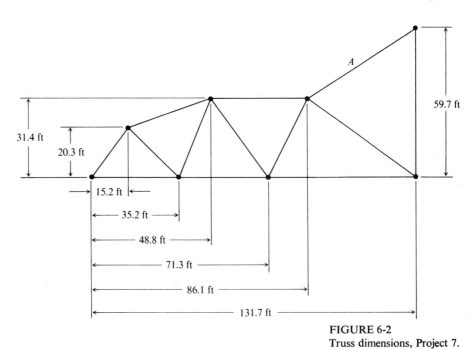

FIGURE 6-1
Derrick in Project 6.

FIGURE 6-2
Truss dimensions, Project 7.

where q = total heat flow, Btu/h

Δt = temperature difference across entire wall, °F

L = thickness of single layer, ft

k = thermal conductivity of layer, Btu/ft²h °F/ft

A = cross-section area, ft²

and the temperature drop across each layer is

$$t = q \frac{L}{kA} = qR$$

where R is called the thermal resistance of that layer.

Write a program that will compute the total heat flow through a wall of any number of layers, given the total temperature drop across the wall, and L, k, and A for each layer.   Also compute the temperature at the interfaces between layers.

Run your program for a four-layer wall, where one side of the wall is at 80°F and the other side is at 0°F, A (for all layers) is 100 ft², and the conductivities and thicknesses are, starting with the warmer side,

Layer	1	2	3	4
k	0.632	0.068	1.39	0.337
L (ft)	0.578	1.06	0.358	0.783

Ans. $q = 419.2$ Btu/h

$t_1 = 76.17$°F

$t_2 = 10.82$°F

$t_3 = 9.74$°F

12   A steel firebox wall is 1 in. thick and has one surface kept at 1345°F by the burning of fuel, and the other side is kept at 65°F by being in contact with circulating water.   A layer of soot builds up on the flame side at a rate of 0.02 in./year while a layer of boiler scale builds up on the wet side at the rate of 0.01 in./year.   Compute the heat flow rate, in Btu/h, through 1 ft² of the wall when the furnace is new (no soot or scale), and after each succeeding year for a total of 20 years.   Plot your results.   Take the conductivities of steel, soot, and scale as 26.1, 0.067, and 0.043 Btu/ft²h °F/ft, respectively. Refer to Project 11 for the necessary relationships.

Check   The heat flow rate after 10 years = 2872 Btu/h

13   Write a program to compute the total resistance of any number of resistors, connected in parallel.   What will be the resistance of the parallel combination of 142, 99.4, 45.2, and 73.9 Ω?   *Ans.*   18.9544 Ω

14   Write a program that will print the inductive reactance, the capacitive reactance, the total impedance, phase angle, and resonant frequency of any circuit having a capacitance of C microfarads, an inductance of L millihenrys, and a resistance of R ohms and is

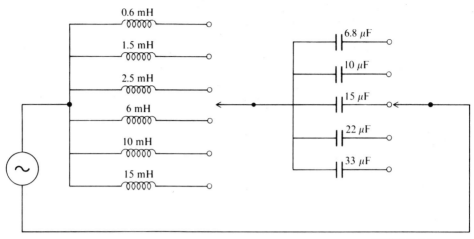

FIGURE 6-3
Circuit for Project 16.

subjected to an alternating voltage at a frequency of f hertz.   Test the program when
$R = 15$, $L = 4.61$, $C = 12.4$, and $f = 450$.   *Ans.* $X_C = -28.52\ \Omega$
$$X_L = 13.03\ \Omega$$
$$X = -15.49\ \Omega$$
$$Z = 21.56\ \Omega$$
$$\phi = -45.92°$$
$$f = 665.67\ \text{Hz}$$

15 Write a program that will compute the total impedance and phase angle for any *RLC*
circuit, for source frequencies from 0 to 1000 Hz in steps of 50 Hz.   Run your program
with the circuit values given in Project 14, and plot the resulting points.
Check   At 500 Hz, $X = 18.71$, $\phi = -36.72°$

16 A circuit in a certain electronic device contains five capacitors and six inductors which
may be independently selected, as in Fig. 6-3.   An alternating voltage is applied to the
circuit, and it is desired to choose a source frequency that will not cause resonance in
the circuit.   Write a program that will compute the resonant frequency in hertz for the
30 possible capacitor-inductor combinations.
Check $F = 438$ Hz when $C = 22\ \mu$F and $h = 6$ mH

17 Most periodic functions may be expressed as the sum of an infinite number of sinusoidal
waves of different frequencies (a Fourier series).   For a square wave of unit amplitude,
the ordinate y at any value of x is given as

$$y = \frac{4}{\pi}\left(\sin x + \frac{1}{3}\sin 3x + \frac{1}{5}\sin 5x + \cdots\right)$$

Show that this series will indeed generate a square wave by computing y for values of x from 0 to 2π in steps of $\pi/10$.   Compute each y, using the first 50 terms of the Fourier series.

18   A square wave with an amplitude of 1 V and a frequency of 60 Hz is applied to the terminals of a pure inductance of 10 mH.   Compute the amplitude of the resulting current wave at 20 intervals over a single cycle of the square wave.   Use the first 10 terms of the Fourier expansion for the square wave given in Project 17, replacing x by ωt, where ω is the frequency in radians per second and t is elapsed time.

   Hint   At each instant of time, compute the current due to each of the components of the square wave (up to 10) and sum them to get the total current at that instant.   Remember that the reactance of the circuit will be different at each frequency.

   Check   The current is 0.301 A at 5 ms

PART TWO

Mathematical Methods for Computers

7

SYSTEMS OF LINEAR EQUATIONS

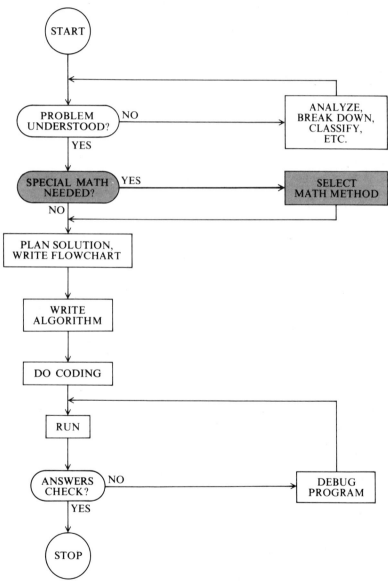

FIGURE 7-1
Steps to follow in solving a complex problem.

We shall see in Chap. 9, in applying the method of least squares, we will be required to find the solution of a system of linear equations. Such systems arise in many problems, particularly in electrical-network computations and in the solution of partial differential equations. To solve such a system means to find a set of values of the variables that will *simultaneously* satisfy all the equations.

The systems usually encountered in engineering are those in which the number of equations just equals the number of unknowns. These are the only type that will be considered here.

7-1 SYSTEMS OF TWO OR THREE EQUATIONS

Systems containing only two or three equations (and unknowns) can readily be solved for all unknowns by the techniques of addition and subtraction, substitution, comparison, or by determinants, as described in most texts on college algebra. For two linear equations in two variables

$$a_1 x + b_1 y = k_1$$
$$a_2 x + b_2 y = k_2 \tag{7-1}$$

we get

$$x = \frac{b_2 k_1 - b_1 k_2}{a_1 b_2 - a_2 b_1} \tag{7-2}$$

$$y = \frac{a_1 k_2 - a_2 k_1}{a_1 b_2 - a_2 b_1} \tag{7-3}$$

Obviously, for a solution to exist, the denominator $a_1 b_2 - a_2 b_1$ must not be zero.

For three linear equations in three variables

$$a_1 x + b_1 y + c_1 z = k_1$$
$$a_2 x + b_2 y + c_2 z = k_2$$
$$a_3 x + b_3 y + c_3 z = k_3 \tag{7-4}$$

we get, by any of the previously mentioned methods,

$$x = \frac{b_2 c_3 k_1 + b_1 c_2 k_3 + b_3 c_1 k_2 - b_2 c_1 k_3 - b_3 c_2 k_1 - b_1 c_3 k_2}{\Delta_3}$$

$$y = \frac{a_1 c_3 k_2 + a_3 c_2 k_1 + a_2 c_1 k_3 - a_3 c_1 k_2 - a_1 c_2 k_3 - a_2 c_3 k_1}{\Delta_3} \tag{7-5}$$

$$z = \frac{a_1 b_2 k_3 + a_3 b_1 k_2 + a_2 b_3 k_1 - a_3 b_2 k_1 - a_1 b_3 k_2 - a_2 b_1 k_3}{\Delta_3}$$

where

$$\Delta_3 = a_1 b_2 c_3 + a_3 b_1 c_2 + a_2 b_3 c_1 - a_3 b_2 c_1 - a_1 b_3 c_2 - a_2 b_1 c_3 \qquad (7\text{-}6)$$

Again, the denominator Δ_3 must not equal zero.

These equations are easily programmed. In each case the denominator should be examined, and if it is zero the run should be terminated.

7-2 ITERATION TECHNIQUES

Many iteration techniques exist for solving various types of problems, but they all have the same general features. These are:

1 *Guess the answer* to the problem. The first guess should be as close to the true answer as you can estimate and can be aided by your knowledge of the physical situation. Usually you can at least estimate the order of magnitude of the answer and whether it is positive or negative.

2 Using this first guess as a starting point, obtain a *closer approximation* to the answer by means of a suitable numerical technique.

3 Using the second approximation obtained in step 2, apply the technique again to obtain a *third approximation* to the answer.

4 Continue to obtain *successive approximations* until the desired accuracy is obtained.

If each successive approximation is closer to the correct answer than the previous approximation, and if the differences between approximations get smaller and smaller, the iteration is said to *converge*. If the iteration does converge, you may watch the differences between successive approximations and halt the run when this difference is smaller than the least significant figure required. This will work well except where the iteration converges very slowly. This will cause small differences in successive figures while you may still be a long way from the correct answer.

Not all iterations will converge; they may *diverge* instead. Not all iterations that do converge will converge fast enough to be useful.

7-3 AN ITERATION TECHNIQUE: THE GAUSS-SEIDEL METHOD

Iterative solutions are well suited to computer use. One of the several techniques available for solving sets of equations is the Gauss-Seidel method. This method can be used on systems having any number of equations, but it is not guaranteed to converge in all cases.

To use the method, first solve each equation in turn for one of the variables in terms of the remaining variables. Thus the system (7-4), for example, would be written

$$x = \frac{k_1 - b_1 y - c_1 z}{a_1} \tag{7-7}$$

$$y = \frac{k_2 - a_2 x - c_2 z}{b_2} \tag{7-8}$$

$$z = \frac{k_3 - a_3 x - b_3 y}{c_3} \tag{7-9}$$

Next, guess the values of y and z as closely as you can. In an electrical-network problem, for example, you may at least know the order of magnitude and sign of the currents and voltages to be expected. If you have no such information, then take any values for y and z, say 0,0. These values are substituted into Eq. (7-7) and x is computed. Using this value of x and your initial guess for z, solve Eq. (7-8) for y. Using this value for y and your computed value for x, solve Eq. (7-9) for z. These three computed values are (one hopes) closer to the true solution than were the first guesses. Using these values, substitute in the equations again, computing a new set of answers, and always using the latest value of a variable as soon as it is computed. Continue iterating until the change in variables is smaller than the required accuracy. In the case of divergence, stop when you have gone through the loop a given number of times. To guard against being fooled by a slowly converging iteration, you should check your answers by substituting them back into the original equations. Should you fail to converge on a solution, try rearranging the order of the equations.

EXAMPLE 7-1   Solve the set of equations

$$3x + y = 5$$
$$2y - 3z = -5$$
$$x + 2z = 7$$

using the Gauss-Seidel method.

SOLUTION   Solving the three equations for x, y, and z, respectively, we get

$$x = \frac{5 - y}{3}$$

$$y = \frac{3z - 5}{2}$$

$$z = \frac{7 - x}{2}$$

Not having any ideas as to the size of the answers, let us take our first guess as $y = z = 0$, from which we compute

$$x = \frac{5 - 0}{3} = \frac{5}{3}$$

Letting z equal zero in the second equation, we get

$$y = \frac{0 - 5}{2} = \frac{-5}{2}$$

Letting x equal $\frac{5}{3}$ in the third equation, we get

$$z = \frac{7 - \frac{5}{3}}{2} = \frac{8}{3}$$

Letting y equal $-\frac{5}{2}$ and z equal $\frac{8}{3}$, we return to the first equation and recompute x, and so on.   The remainder of this computation has been done by computer, with the values of x, y, and z being printed after each pass through the three equations; the results are as follows;

x	y	z
0	0	0
1.66667	−2.5	2.66667
2.5	1.5	2.25
1.16667	0.875	2.91667
1.375	1.875	2.8125
1.04167	1.71875	2.97917
1.09375	1.96875	2.95312
1.01042	1.92969	2.99479
1.02344	1.99219	2.98828
1.0026	1.98242	2.9987
1.00586	1.99805	2.99707
1.00065	1.99561	2.99967
1.00146	1.99951	2.99927
1.00016	1.9989	2.99992
1.00037	1.99988	2.99982
1.00004	1.99973	2.99998
1.00009	1.99997	2.99995
1.00001	1.99993	2.99999
1.00002	1.99999	2.99999
1.	1.99998	3.
1.00001	2.	3.
1.	2.	3.
1.	2.	3.
1.	2.	3.

It took 21 iterations to converge upon a solution.   We should not neglect to check our work by substituting our solution back into the original equations.   Doing this, we get

$$3(1) + 2 - 5 = 0$$
$$2(2) - 3(3) + 5 = 0$$
$$1 + 2(3) - 7 = 0 \qquad \text{////}$$

7-4 GAUSS ELIMINATION

If the Gauss-Seidel method fails to converge, even after rearranging the equations, you may resort to the method of elimination. It is quite simple in concept and is often the method taught in an algebra course. It is somewhat more difficult to program than the Gauss-Seidel method because it requires the clever use of doubly subscripted variables. However, you have to struggle through the program writing only once and can carefully save it for future use.

Before looking at the programming, let us first solve a set of equations by this method. Take the system

$$2x + y - z = 5$$
$$3x - 2y + 2z = -3$$
$$x - 3y - 3z = -2$$

We may perform the following operations on any set of equations without changing the solutions:

1 Multiply an equation by any number but zero.
2 Write the equations in a different order.
3 Add a multiple of any equation to any other equation.

Let us apply these operations to our system.   First, let us divide each equation by the coefficient of the *x* term in that equation.   We get

$$x + \frac{y}{2} - \frac{z}{2} = \frac{5}{2}$$

$$x - \frac{2y}{3} + \frac{2z}{3} = -1$$

$$x - 3y - 3z = -2$$

We now subtract the first equation from the second and third:

$$x + \frac{y}{2} - \frac{z}{2} = \frac{5}{2}$$

$$-\frac{7y}{6} + \frac{7z}{6} = -\frac{7}{2}$$

$$-\frac{7y}{2} - \frac{5z}{2} = -\frac{9}{2}$$

Again we divide each equation by the coefficient of its first term:

$$x + \frac{y}{2} - \frac{z}{2} = \frac{5}{2}$$

$$y - z = 3$$

$$y + \frac{5z}{7} = \frac{9}{7}$$

Subtracting the second equation from the third:

$$x + \frac{y}{2} - \frac{z}{2} = \frac{5}{2}$$

$$y - z = 3$$

$$\frac{12z}{7} = -\frac{12}{7}$$

Notice that we have *eliminated* all unknowns but one from the last equation. Our system is now said to be in *triangular form*, and we can easily solve for the three unknowns.

$$z = (-\tfrac{12}{7})(\tfrac{7}{12}) = -1$$
$$y = 3 + (-1) = 2$$
$$x = \tfrac{5}{2} + \tfrac{1}{2}(-1) - \tfrac{1}{2}(2) = 1$$

Let us now see how we can instruct a computer to carry out the same operations. Our task will be made easier if we write our equations in the following form:

$$a_{11}x_1 + a_{12}x_2 + a_{13}x_3 = a_{14}$$
$$a_{21}x_1 + a_{22}x_2 + a_{23}x_3 = a_{24}$$
$$a_{31}x_1 + a_{32}x_2 + a_{33}x_3 = a_{34}$$

From the previous example, it is obvious that we can solve the equations just as well without having to write in the x's, since their positions in the equations are always the same. Let us rewrite the equations leaving out the x's and also omitting all plus and equal signs.

$$
\begin{array}{cccc}
a_{11} & a_{12} & a_{13} & a_{14} \\
a_{21} & a_{22} & a_{23} & a_{24} \\
a_{31} & a_{32} & a_{33} & a_{34}
\end{array}
$$

We are left with an *array* containing all the information given in the original equations, but in a much more compact form. This particular array has three rows and four columns, and the subscripts for each *element* identify first the row in which it is located and then the column. We shall set aside 12 locations in the computer memory to store these 12 numbers, having addresses $A(1,1)$, $A(1,2)$, ..., $A(P,J)$, ..., $A(3,4)$, where P is the row number and J the column number. At this point you may want to glance back to Sec. 2-9 to refresh your memory as to how we used doubly subscripted variables to represent an array, or table.

Our task is then to manipulate this array into triangular form so that all numbers but the last on the *main diagonal* (a_{11}, a_{22}) equal 1, and the numbers below the main diagonal (a_{21}, a_{31}, a_{32}) all become zeros. To accomplish this we shall follow the same steps used in solving the three equations in the previous example.

Our first step was to divide each coefficient by the first coefficient in each row. We may do this with the following statements:

FOR P = 1 TO 3
FOR J = 1 TO 4
LET A(P,J) = A(P,J)/A(P,1)
NEXT J
NEXT P

Because of its simplicity and clarity, we have written the nested loops as they would appear in Basic, as shown in Sec. 2-7.[1] Trace through the above statements to verify that they will result in the following array:

$$\begin{array}{cccc} 1 & a_{12} & a_{13} & a_{14} \\ 1 & a_{22} & a_{23} & a_{24} \\ 1 & a_{32} & a_{33} & a_{34} \end{array}$$

Keep in mind that the numerical values of each of the a's are not necessarily the same as they were in the original array, even though we still denote them by the same symbol. We are merely putting a different number into the same memory location.

---

[1] This algorithm written without the special loop instructions such as the FOR and NEXT statements in Basic and the DO statement in Fortran would be quite messy and hard to read. The FOR and NEXT statements are easy to understand even if you do not know the Basic language. For example, in the program

```
10   FOR X = 3 TO 7
20   PRINT X
30   NEXT X
40   END
```

the variable X is given an initial value of 3 in line 10; this 3 is printed in line 20, and line 30 sends the computer back to line 10. The value of X is then increased by 1 and line 20 causes the number 4 to be printed. Line 30 again sends the computer back to line 10. Looping is continued in this way until the number 7 is reached and printed. At line 30 the computer will recognize that another increase in X will put it above its upper limit and so does not return to line 10 but proceeds to the line following the NEXT (40 in this case). Thus this program causes the numbers 3, 4, 5, 6, 7 to be printed.

Our next step is to subtract the first row from the second and third rows; this may be done with the statements

```
FOR P = 2 TO 3
FOR J = 1 TO 4
LET A(P,J) = A(P,J) − A(1,J)
NEXT J
NEXT P
```

and our array becomes

$$
\begin{matrix}
1 & a_{12} & a_{13} & a_{14} \\
0 & a_{22} & a_{23} & a_{24} \\
0 & a_{32} & a_{33} & a_{34}
\end{matrix}
$$

We now want to divide the second row by a_{22} and the third row by a_{32}. We can do this with the same statements we used before, except that we wish to start with the second row and second column, rather than the first row and first column as before.

```
FOR P = 2 TO 3
FOR J = 2 TO 4
LET A(P,J) = A(P,J)/A(P,2)
NEXT J
NEXT P
```

Notice that the only change we have made in these statements is that all the 1s in the previous statements have been replaced by 2s. This gives us the idea that we could use a single set of statements for both computations such as

```
FOR P = K TO 3
FOR J = K TO 4
LET A(P,J) = A(P,J)/A(P,K)
NEXT J
NEXT P
```

where K is set to 1 the first time we use this loop, and to 2 for the second use. Our array now becomes

$$
\begin{matrix}
1 & a_{12} & a_{13} & a_{14} \\
0 & 1 & a_{23} & a_{24} \\
0 & 1 & a_{33} & a_{34}
\end{matrix}
$$

We now subtract the second equation from the third. Again, we can use the same statements we used before, with the following modifications:

```
FOR P = K + 1 TO 3
FOR J = K TO 4
LET  A(P,J) = A(P,J) − A(K,J)
NEXT J
NEXT P
```

and our array is now in the form

$$
\begin{array}{cccc}
1 & a_{12} & a_{13} & a_{14} \\
0 & 1 & a_{23} & a_{24} \\
0 & 0 & a_{33} & a_{34}
\end{array}
$$

Dividing the third row by its leading coefficient, we get the triangular form:

$$
\begin{array}{cccc}
1 & a_{12} & a_{13} & a_{14} \\
0 & 1 & a_{23} & a_{24} \\
0 & 0 & 1 & a_{34}
\end{array}
$$

The instructions used to put the array into triangular form are summarized here. Notice that where we had the number 3 (which represented the number of equations) in our previous statements, we how have the symbol N, and where we had the number 4 we now have $N + 1$.   With this slight change we can now use these instructions for a system containing any number of equations.

```
FOR  K = 1 TO N
  FOR P = K TO N
    LET  D = A(P,K)
      FOR J = K TO N + 1
        LET  A(P,J) = A(P,J)/D
      NEXT J
  NEXT P
  FOR P = K + 1 TO N
    FOR J = K TO N + 1
        LET  A(P,J) = A(P,J) − A(K,J)
    NEXT J
  NEXT P
NEXT K
```

We can now compute the values of x_1, x_2, and x_3.

$$
\begin{aligned}
x_3 &= a_{34} \\
x_2 &= a_{24} - a_{23}x_3 \\
x_1 &= a_{14} - a_{13}x_3 - a_{12}x_2
\end{aligned}
$$

The statements needed to compute and print the answers are given in the last section of the complete program which follows.   The first section reads N and all the coefficients into the memory, and the middle section is identical to the one given above for putting the array into triangular form.   Notice the use of indenting and spacing to separate the various loops and make the program easier to read.

```
READ-IN
    READ N
    FOR P = 1 TO N
        FOR J = 1 TO N + 1
            READ A(P,J)
        NEXT J
    NEXT P

TRIANGULARIZATION
    FOR K = 1 TO N
        FOR P = K TO N
            LET D = A(P,K)
            FOR J = K TO N + 1
                LET A(P,J) = A(P,J)/D
            NEXT J
        NEXT P

        FOR P = K + 1 TO N
            FOR J = K TO N + 1
                LET A(P,J) = A(P,J) − A(K,J)
            NEXT J
        NEXT P
    NEXT K

BACK SUBSTITUTION
and PRINTOUT
    FOR K = N TO 1 STEP −1
        LET X(K) = A(K, N + 1)
        FOR W = N TO K STEP −1
            LET X(K) = X(K) − A(K, W + 1) * X(W + 1)
        NEXT W
        PRINT X(K)
    NEXT K
```

This program also requires a data statement, giving first the number of equations N and then the numbers in the array given by row

$A(1,1), A(1,2), A(1,3), \ldots, A(3,4)$

This program will not work if, in the eleventh line, we try to divide by a D which is zero.   Should this occur, rearrange the order of the equations and try again.

If you are solving a system in which many of the a's are zero (a "sparse" system) it may not be possible to avoid division by zero, and the Gauss-Seidel method should be tried. Fortunately the Gauss-Seidel method works better for sparse systems.

EXERCISE 7-1

1 Solve the following sets of equations for x and y.

(a) $3x - 2y = 1$
$5x + y = 6$

Ans. $x = 1$
$y = 1$

(b) $2x + 5y = 8$
$3x - 2y = -7$

Ans. $x = -1$
$y = 2$

(c) $x - 14y = -1$
$2x + 7y = 3$

Ans. $x = 1$
$y = \frac{1}{7}$

(d) $12x + 5y = 0$
$8x + 10y = 8$

Ans. $x = -\frac{1}{2}$
$y = 1.2$

2 Solve the following sets of equations for x, y, and z.

(a) $x + 2y - z = -3$
$3x + y + z = 4$
$x - y + 2z = 6$

Ans. $x = 1$, $y = -1$, $z = 2$

(b) $2x + 4y + z = 5$
$x + y + z = 6$
$2x + 3y + z = 6$

Ans. $x = 2$, $y = -1$, $z = 5$

(c) $x - y = 5$
$y - z = -6$
$2x - z = 2$

Ans. $x = 3$, $y = -2$, $z = 4$

(d) $5x - 2y - z = 0$
$7x + 4y + 2z = 0$
$6x + 6y - 2z = 5$

Ans. $x = 0$, $y = \frac{1}{2}$, $z = -1$

3 Solve the following sets of equations for x, y, z, and w.

(a) $x - 2y - 3z + w = 7$
$-x + 3y + 5z - w = -9$
$2x + y - z + 3w = 18$
$3x - y + z + 4w = 15$

Ans. $x = 1$, $y = 2$, $z = -2$, $w = 4$

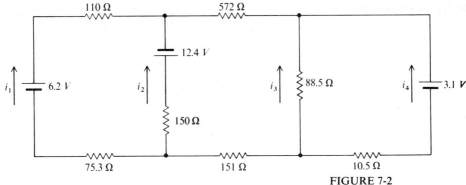

FIGURE 7-2
Three-loop electrical network.

(b) $2x + y + 2z - w = 0$
$6x + 8y + 12z - 13w = -21$
$10x + 2y + 2z + 3w = 21$
$-4x + z - 3w = -13$ *Ans.* $x = \frac{3}{2}, \quad y = 1, \quad z = -1, \quad w = 2$

PROJECTS

1 Apply Kirchhoff's laws to the circuit of Fig. 7-2 and obtain a set of four equations. Solve these equations simultaneously by any of the methods in this chapter to obtain the values of the four currents shown.

Ans. $i_1 = 52.1$ mA, $i_2 = -60.4$ mA,
$i_3 = -30.4$ mA, $i_4 = 38.8$ mA

8

INTERPOLATION

8-1 EQUATIONS AND TABULAR DATA

If the equation relating one variable y with another variable x is known, it is usually possible to perform any of the common mathematical operations, such as finding roots, rate of change of the function, area under the curve, etc. Suppose, however, that the functional relation is not known, as is often the case in practical engineering problems. Technical information very often comes in the form of a table of data, where one variable has been measured as another variable has been changed.

Suppose, for example, you wanted to measure the acceleration characteristics of your car. You might take it to a straight, level stretch of road, preferably free of traffic, cattle, and state troopers, bringing with you a stopwatch and a friend to read it so that you can keep your eyes on the road. Starting from rest, you might then accelerate, taking speedometer readings every 2 s or so. The result would be a table of data similar to Table 8-1; these data points, when plotted, would appear as in Fig. 8-1.

Now these data are fine as far as they go, but the value of your experiment would be greatly increased if you could deduce more information from the original data. For instance, can you find the speed at *any* instant of time, rather than only at the tabulated times? Also, how far has the car traveled at any instant? What is the acceleration at any instant? How long does it take the car to reach, say, 40 mi/h? Many

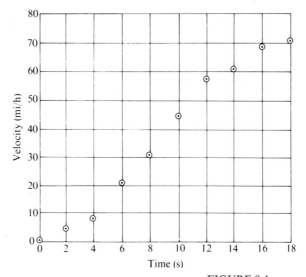

FIGURE 8-1
Plot of speedometer readings vs. time.

more such questions could be asked.   They could readily be answered if you had an equation relating the velocity to time; differentiation of this equation would give acceleration, and integration of the same equation would give displacements.

But you *do not have* an equation!

In practice, such data are often handled by graphical methods.   The data are plotted and a smooth curve drawn through the points.   Intermediate values between data points are obtained simply by reading the coordinates of any desired point on the drawn curve.   Roots are read off the graph where the curve crosses the horizontal axis.   Tangents are drawn to the curve to obtain the slopes at any points, and integration is performed by measuring the area under the curve with a planimeter, or by "counting boxes."

Table 8-1   CAR SPEED VS. TIME

Time, s	Speed, mi/h	Time, s	Speed, mi/h
0	0	10	45
2	4	12	57
4	8	14	61
6	21	16	69
8	31	18	71

Numerical methods, suitable for programming on a computer, have been developed to handle such tabular data. We may approach the problem in two alternative ways.

We may choose to find an equation relating the two variables in the table over the entire range of data. The equation usually sought is one that will "smooth" the given data; such data are often corrupted by the presence of random measuring errors or by "noise" (random fluctuations in the variables themselves). Once such an equation (called an *empirical formula*) has been found, the usual mathematical techniques can be applied. This is the general problem of *curve fitting* described in the following chapter.

Instead of fitting an equation to the entire data, we may attempt the easier task of fitting an equation to just a *small portion* of the data. We would, in effect, be chopping up the data into small sections, each containing the same number of points (usually two, three, or four). We then write an equation connecting the points in one such section, use this equation to perform any necessary computations, and then repeat the procedure for the following section.

The process of writing an equation passing through a small number of data points is called *interpolation* and is described in the remainder of this chapter.

8-2 LINEAR INTERPOLATION

Interpolation is not new to you for you have often used linear interpolation to find intermediate values in tables of mathematical functions. What you did, in effect, was to approximate the tabulated function by a straight line passing through the two tabulated points on either side of the missing point. The equation of this line was then solved for the desired ordinate, y.

Referring to Fig. 8-2, the equation of the line connecting the two tabulated points P_1 and P_2 is

$$\frac{y - y_0}{x - x_0} = \frac{y_1 - y_0}{x_1 - x_0}$$

If we let s be the fractional part of the interval between the two given values of x so that

$$s = \frac{x - x_0}{x_1 - x_0}$$

we get

$$y = y_0 + s(y_1 - y_0) \tag{8-1}$$

The simple flowchart for performing linear interpolation is given in Fig. 8-3.

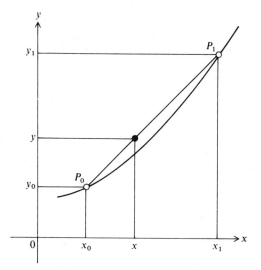

FIGURE 8-2
Linear interpolation.

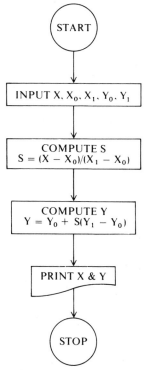

FIGURE 8-3
Flowchart for linear interpolation.

EXAMPLE 8-1   Referring to Table 8-1, find the velocity at 7.2 s by linear interpolation in the table.

The fractional part of the interval is

$$s = \frac{7.2 - 6}{8 - 6} = 0.6$$

and so

$$V = 21 + 0.6(31 - 21) = 27 \qquad\qquad ////$$

EXAMPLE 8-2   Use linear interpolation in Table 8-2 to find the sine of 14.186°. The fractional part of the interval will be

$$s = \frac{14.186 - 14.1}{0.1} = 0.86$$

and so

$$\sin \theta = 0.24362 + 0.86(0.24531 - 0.24362) = 0.2450734$$

Since the original table contained only five significant figures, we round our answer to $y = 0.24507$. $\qquad\qquad ////$

8-3   POLYNOMIAL INTERPOLATION

If higher accuracy than can be obtained by linear interpolation is desired, an equation may be written connecting *any* number of points in the region of interest; the more points used, the better (usually) the accuracy.   The type of function chosen is usually a polynomial, although other functions such as a Fourier series could also be used.

Table 8-2   SIN θ FOR DECIMAL FRACTIONS OF A DEGREE

θ, deg	$\sin \theta$
14.0	0.24192
14.1	0.24362
14.2	0.24531
14.3	0.24700
14.4	0.24869

One widely used interpolating polynomial is the *Lagrange interpolation polynomial*. This formula will be discussed in the following two sections for the general case of unequally spaced intervals of the independent variable x and then for the more usual case of equally spaced intervals.

8-4 LAGRANGE'S INTERPOLATION POLYNOMIAL FOR UNEQUAL INTERVALS

Given a set of $n + 1$ tabulated points

$$(x_0, y_0), (x_1, y_1), (x_2, y_2), \ldots, (x_n, y_n)$$

we seek an equation of a curve that will pass through all these points. One such equation is the Lagrange polynomial:

$$y = L_0(x)y_0 + L_1(x)y_1 + L_2(x)y_3 + \cdots + L_n(x)y_n \tag{8-2}$$

where

$$L_0(x) = \frac{(x - x_1)(x - x_2)(x - x_3) \cdots (x - x_n)}{(x_0 - x_1)(x_0 - x_2)(x_0 - x_3) \cdots (x_0 - x_n)}$$

$$L_1(x) = \frac{(x - x_0)(x - x_2)(x - x_3) \cdots (x - x_n)}{(x_1 - x_0)(x_1 - x_2)(x_1 - x_3) \cdots (x_1 - x_n)}$$

$$L_2(x) = \frac{(x - x_0)(x - x_1)(x - x_3) \cdots (x - x_n)}{(x_2 - x_0)(x_2 - x_1)(x_2 - x_3) \cdots (x_2 - x_n)} \tag{8-3}$$

$$\cdots\cdots\cdots\cdots\cdots\cdots\cdots\cdots\cdots$$

$$L_n(x) = \frac{(x - x_0)(x - x_1)(x - x_2) \cdots (x - x_{n-1})}{(x_n - x_0)(x_n - x_1)(x_n - x_2) \cdots (x_n - x_{n-1})}$$

At first glance this equation looks too complicated for anyone to ever want to use, but it will not be so bad after we simplify it a bit for a specific number of data points. We will do this later on, but first let us see if this equation really connects all our data points.

We may convince ourselves that this polynomial passes through each of our data points by evaluating it at the point (x_r, y_r). Substituting x_r for x in the first of Eqs. (8-3) we get

$$L_0(x_r) = \frac{(x_r - x_1)(x_r - x_2) \cdots (x_r - x_r) \cdots (x_r - x_n)}{(x_0 - x_1)(x_0 - x_2) \cdots (x_0 - x_r) \cdots (x_0 - x_n)} = 0$$

In a similar way, all the other L's will vanish, except

$$L_r(x_r) = \frac{(x_r - x_0)(x_r - x_1) \cdots (x_r - x_n)}{(x_r - x_0)(x_r - x_1) \cdots (x_r - x_n)} = 1$$

Therefore, Eq. (8-2) reduces to

$$y = L_r(x_r)y_r = y_r$$

showing that the point (x_r, y_r) does in fact satisfy the equation.

EXAMPLE 8-3   Write the equation of a curve that will pass through the four points (1,1) (3,4) (4,2), and (6,5).   Thus

$$
\begin{aligned}
x_0 &= 1 & y_0 &= 1 \\
x_1 &= 3 & y_1 &= 4 \\
x_2 &= 4 & y_2 &= 2 \\
x_3 &= 6 & y_3 &= 5
\end{aligned}
$$

From Eqs. (8-3),

$$L_0(x) = \frac{(x-3)(x-4)(x-6)}{(1-3)(1-4)(1-6)} = \frac{x^3 - 13x^2 + 54x - 72}{-30}$$

$$L_1(x) = \frac{(x-1)(x-4)(x-6)}{(3-1)(3-4)(3-6)} = \frac{x^3 - 11x^2 + 34x - 24}{6}$$

$$L_2(x) = \frac{(x-1)(x-3)(x-6)}{(4-1)(4-3)(4-6)} = \frac{x^3 - 10x^2 + 27x - 18}{-6}$$

$$L_3(x) = \frac{(x-1)(x-3)(x-4)}{(6-1)(6-3)(6-4)} = \frac{x^3 - 8x^2 + 19x - 12}{30}$$

Substituting in Eq. (8-2) we get

$$y = \frac{(x^3 - 13x^2 + 54x - 72)1}{-30} + \frac{(x^3 - 11x^2 + 34x - 24)4}{6}$$

$$+ \frac{(x^3 - 10x^2 + 27x - 18)2}{-6} + \frac{(x^3 - 8x^2 + 19x - 12)5}{30}$$

which reduces to

$$y = \tfrac{1}{30}(14x^3 - 147x^2 + 451x - 288)$$

which is the equation we seek.

CHECK   Does this equation pass through all four given points?   We can see if each of the original points satisfy the equation.
When $x = 1$

$$y = \tfrac{1}{30}(14 - 147 + 451 - 288) = \tfrac{30}{30} = 1$$

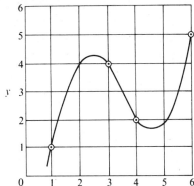

FIGURE 8-4
Plot of interpolating polynomial of x
Example 8-3.

When $x = 3$

$$y = \tfrac{1}{30}[14(27) - 147(9) + 451(3) - 288] = \tfrac{120}{30} = 4$$

When $x = 4$

$$y = \tfrac{1}{30}[14(64) - 147(16) + 451(4) - 288] = \tfrac{60}{30} = 2$$

When $x = 6$

$$y = \tfrac{1}{30}[14(216) - 147(36) + 451(6) - 288] = \tfrac{150}{30} = 5$$

We see that our equation is indeed satisfied by all four given points.   This equation and the given points are plotted in Fig. 8-4.

If we now desired the value of y at some other value of x in this region, we would merely insert it into our equation and solve for y.   For example, for $x = 3.428$, we get

$$y = \tfrac{1}{30}[14(3.428)^3 - 147(3.428)^2 + 451(3.428) - 288] = 3.152 \qquad \text{////}$$

8-5 LAGRANGE'S POLYNOMIAL SIMPLIFIED FOR THREE-POINT INTERPOLATION

The Lagrange polynomial is rather forbidding in the general form of Eqs. (8-2) and (8-3) and is also inconvenient to program in that form.   Let us simplify these equations for the case of three unequally spaced data points (x_0, y_0), (x_1, y_1), and (x_2, y_2).

Interpolation over three data points is sometimes called *quadratic interpolation* because the interpolating polynomial will turn out to be of second degree.   Let

$$R_0 = \frac{y_0}{(x_0 - x_1)(x_0 - x_2)}$$

$$R_1 = \frac{y_1}{(x_1 - x_0)(x_1 - x_2)}$$

$$R_2 = \frac{y_2}{(x_2 - x_0)(x_2 - x_1)}$$

Then, from Eqs. (8-3),

$$L_0(x)y_0 = R_0(x - x_1)(x - x_2)$$
$$L_1(x)y_1 = R_1(x - x_0)(x - x_2)$$
$$L_2(x)y_2 = R_2(x - x_0)(x - x_1)$$

Equation (8-2) then becomes

$$y = R_0(x - x_1)(x - x_2) + R_1(x - x_0)(x - x_2) + R_3(x - x_0)(x - x_1)$$
$$= R_1[x^2 - (x_1 + x_2)x + x_1 x_2] + R_1[x^2 - (x_0 + x_2)x + x_0 x_2]$$
$$+ R_2[x^2 - (x_0 + x_1)x + x_0 x_1]$$

Collecting terms, we get

$$y = C_0 + C_1 x + C_2 x^2$$

where

$$C_0 = R_0 x_1 x_2 + R_1 x_0 x_2 + R_2 x_0 x_1$$
$$C_1 = - R_0(x_1 + x_2) - R_1(x_0 + x_2) - R_2(x_0 + x_1) \qquad (8\text{-}4)$$
$$C_2 = R_0 + R_1 + R_2$$

Equations (8-4) enable us to write an algorithm for three-point interpolation; the flowchart of Fig. 8-5 is such an algorithm.   It finds the second-degree equation passing through the three points (x_0, y_0), (x_1, y_1), and (x_2, y_2) and computes the ordinate Y for some intermediate value of X falling anywhere between x_0 and x_2. Remember that x_0, x_1, and x_2 do not have to be equally spaced.

For the usual case where the x's are *equally spaced* with interval h, our equations simplify to

$$C_0 = \frac{y_0 x_1 x_2 - 2y_1 x_0 x_2 + y_2 x_0 x_1}{2h^2}$$

$$C_1 = \frac{-y_0(x_1 + x_2) + 2y_1(x_0 + x_2) - y_2(x_0 + x_1)}{2h^2} \qquad (8\text{-}5)$$

$$C_2 = \frac{y_0 - 2y_1 + y_2}{2h^2}$$

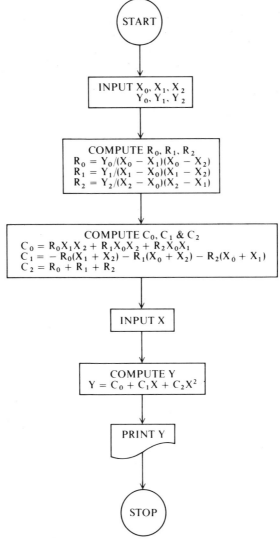

FIGURE 8-5
Flowchart for three-point interpolation.

EXAMPLE 8-4 Referring to Table 8-1, find the velocity at 7.2 s by three-point interpolation in the table. Let

$$
\begin{array}{ll}
S_0 = 6 & V_0 = 21 \\
S_1 = 8 & V_1 = 31 \\
S_2 = 10 & V_2 = 45
\end{array}
$$

Then

$$h = 2$$

and

$$C_0 = \frac{21(8)(10) - 2(31)(6)(10) + 45(6)(8)}{2(4)}$$

$$= \frac{1680 - 3720 + 2160}{8} = 15$$

$$C_1 = \frac{-21(8 + 10) + 2(31)(6 + 10) - 45(6 + 8)}{8}$$

$$= \frac{-378 + 992 - 630}{8} = -2$$

$$C_2 = \frac{21 - 2(31) + 45}{8} = \frac{1}{2}$$

The interpolating polynomial is then

$$V = 15 - 2S + \tfrac{1}{2}S^2$$

When $S = 7.2$

$$V = 15 - 14.4 + \tfrac{1}{2}(51.84) = 26.52$$

slightly lower than the value obtained by linear interpolation.

It is interesting to superimpose our interpolating polynomial upon the original data points of Fig. 8-1, as in Fig. 8-6. Our interpolating polynomial is a parabola and passes precisely through the three chosen data points. It is shown as a solid line in the interpolation region and dashed elsewhere.

Notice that the difference in the values obtained by quadratic and by linear interpolation is smaller than the inaccuracies in the original data, so that nothing was gained by using three-point interpolation in this case. Three- or four-point interpolation is advantageous only where data points are precisely known, as in published tables of mathematical functions or data from carefully performed experiments.

////

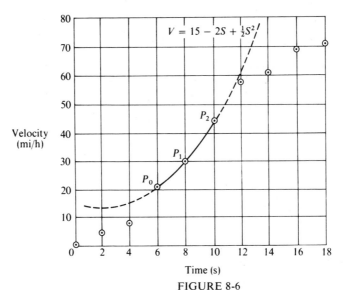

FIGURE 8-6
Plot of interpolating polynomial of Example 8-4.

8-6 FOUR-POINT INTERPOLATION

For still greater accuracy when interpolating in very precise tables, we may choose to use four data points to determine our interpolating polynomial, which will then be of third degree.   Following the same procedure as used in the previous section, we get the equations

$$y = C_0 + C_1 x + C_2 x^2 + C_3 x^3$$

where

$$C_0 = -R_0 x_1 x_2 x_3 - R_1 x_0 x_2 x_3 - R_2 x_0 x_1 x_3 - R_3 x_0 x_1 x_2$$
$$C_1 = R_0(x_1 x_2 + x_1 x_3 + x_2 x_3) + R_1(x_0 x_2 + x_2 x_3 + x_0 x_3)$$
$$\qquad\qquad + R_2(x_0 x_1 + x_1 x_3 + x_0 x_3) + R_3(x_0 x_1 + x_1 x_2 + x_0 x_2)$$
$$C_2 = -R_0(x_1 + x_2 + x_3) - R_1(x_0 + x_2 + x_3) - R_2(x_0 + x_1 + x_3) - R_3(x_0 + x_1 + x_2)$$
$$C_3 = R_0 + R_1 + R_2 + R_3$$
$$R_0 = \frac{y_0}{(x_0 - x_1)(x_0 - x_2)(x_0 - x_3)} \tag{8-6}$$
$$R_1 = \frac{y_1}{(x_1 - x_0)(x_1 - x_2)(x_1 - x_3)}$$

$$R_2 = \frac{y_2}{(x_2 - x_0)(x_2 - x_1)(x_2 - x_3)}$$

$$R_3 = \frac{y_3}{(x_3 - x_0)(x_3 - x_1)(x_3 - x_2)}$$

As with the three-point formulas, simplification is possible when the data points are equally spaced.

EXAMPLE 8-5   An optical filter is placed in the path of a beam of light, absorbing some of the incident energy and transmitting the remainder.   The amount of energy transmitted depends upon its wavelength and is, at each wavelength considered, the product of the incident energy and the percent transmittance of the filter.

The amount of energy in the light beam before filtering and the transmission characteristics of a Wratten No. 58 filter are given as a function of wavelength in Table 8-3.

Interpolating in the table, find the values of incident energy at those wavelengths for which the filter transmission is known.   Multiply the transmittance by the incident energy at each of these wavelengths and print the results in tabular form.

SOLUTION   Let E_W be the energy at wavelength W (thus E_{475} will be the energy at 475 nm or 150 W/μm) and T_W be the transmission at wavelength W.

DATA   We know those values of T_W only for wavelengths that are multiples of 10, and those values of E_W only for wavelengths midway between those for which T_W is known.

UNKNOWNS   These are the values of E_W at wavelengths at which T_W is known and the products of E_W and T_W at each of these wavelengths.   Let us call this product P_W.

RELATIONSHIPS   The product P_W is simply

$$P_W = E_W \cdot \frac{T_W}{100}$$

For finding the intermediate values of E_W we may use Eqs. (8-5) which apply for the case of three equally spaced data points, taking h equal to 100.

OUTPUT FORMAT   Let us print the results in a four-column table containing the wavelength (in increments of 10 nm), the filter transmission, the energy output of the lamp, and the computed product.

PLANNING THE SOLUTION   Let us take as our first three data points (465,135), (475,150) and (485,161), which are the first three tabulated values of E_W with the corresponding values of W.   We can then use these three points in the interpolation formulas to obtain E_{470} and E_{480}.   These values will then be multiplied by the corresponding values of T_W (0.23 and 1.38), and this product is printed, or stored for later printout. We then proceed to the next three data points (485,161), (495,182), and (505,213) and repeat the procedure.   We continue in this fashion until the end of the table is reached. The precise way in which this stepping is accomplished is given in the flowchart in Fig. 8-7.   The results of the actual computation are given in Table 8-4.

Table 8-3   FILTER TRANSMISSION AND LAMP OUTPUT FOR VARIOUS WAVELENGTHS

Wavelength, nm	Percent transmittance	Incident energy, W/μm
460	0	
465		135
470	0.23	
475		150
480	1.38	
485		161
490	4.90	
495		182
500	17.7	
505		213
510	38.8	
515		234
520	52.2	
525		248
530	53.6	
535		271
540	47.8	
545		299
550	38.4	
555		326
560	27.8	
565		348
570	17.4	
575		366
580	9.00	
585		384
590	3.50	
595		422
600	1.50	
605		433
610	0.41	
615		462
620	0	
625		479

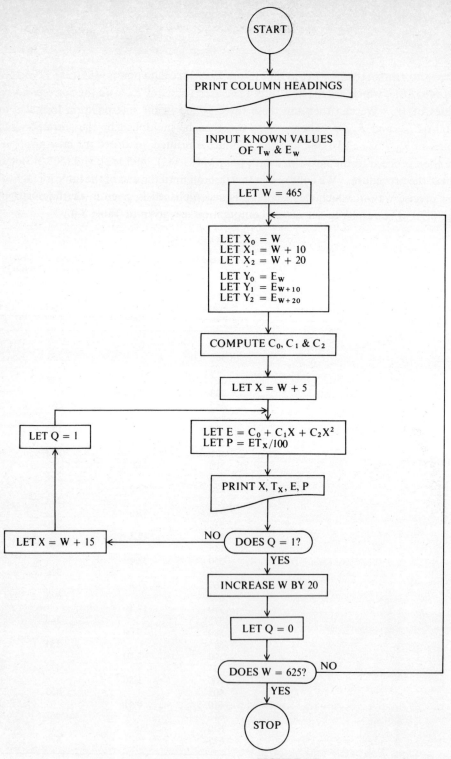

FIGURE 8-7
Flowchart for the solution of Example 8-5.

Table 8-4   RESULTS OF THE COMPUTATION
OF EXAMPLE 8-5

Wavelength	Transmission	Energy	Product
470	0.23	143.0	0.33
480	1.38	156.0	2.15
490	4.90	170.3	8.34
500	17.7	196.3	34.7
510	38.8	224.4	87.1
520	52.2	241.9	126
530	53.6	258.9	139
540	47.8	284.4	136
550	38.4	313.1	120
560	27.8	337.6	93.9
570	17.4	357.0	62.1
580	9.00	375.0	33.8
590	3.50	406.4	14.2
600	1.50	430.9	6.46
610	0.41	449.0	1.84
620	0	472.0	0

CHECKING THE RESULTS   Let us check the answers at some particular wavelength, say 550 nm.   By *linear* interpolation we get

$$E_{550} = \tfrac{1}{2}(326 + 299) = 312.5$$

and

$$P_{550} = \frac{(312.5)(38.4)}{100} = 120$$

which check approximately with our computer results.   ////

EXERCISE 8-1

1   Derive Eqs. (8-5).

2   Derive Eqs. (8-6).

3   Simplify Eqs. (8-6) for the case of equally spaced data points.

4   Using linear interpolation in Table 8-2, find the sine of 14.368°.   *Ans.*   0.24815

5   Using three-point interpolation in Table 8-2, find the sine of 14.368°.   *Ans.*   0.24815

6   Write a program to find the speed at every integral value of time from 0 to 18 s, using three-point interpolation in Table 8-1.   *Check* $V = 38.25$ at 9 s

PROJECTS

1 A cam is to be machined to have the radii shown in the table, for particular values of angle of rotation. Radii are given every 20°.

In order to do the machining, however, the shop requires that radii be given every 5°.

Write a program that will print out the cam radii for 5° increments of angular rotation. Since the equation of the curve is not known, use four-point interpolation between the given values.

Angular rotation, deg	Cam radius, in.
0	1.000
20	1.112
40	1.464
60	2.192
80	3.193
100	3.603
120	3.715
140	3.654
160	3.612
180	4.005
200	4.487
220	4.654
240	4.435
260	3.212
280	1.578
300	1.054
320	1.000
340	1.000
360	1.000

Check   Radius = 3.728 in. at 170°

2 The characteristics of a nonlinear resistor were determined by measuring the current through the resistor for various applied voltages, resulting in the table below. A sinusoidal voltage with a peak-to-peak voltage of 30 V at a frequency of 60 Hz is applied across the resistor. Compute the current through the resistor for applied voltages from −15 to 15 V in steps of 1 V, by three-point interpolation in the table.

Voltage, V	Current, A
−16	−6.52
−12	−3.44
−8	−1.97
−4	−0.846
0	0
4	1.13
8	3.06
12	8.37
16	20.8

Check   Current = 3.72 A at 9 V

9

CURVE FITTING

A great deal of technical information is obtained as the result of laboratory tests, measurements, and observations. Usually one variable is measured while another is changed, with all other quantities being held constant. For example, if you wished to know how the length of a certain wire would change as the applied load changed, you would measure the length under different loadings, keeping constant such things as temperature, wire type and diameter, etc. The result of your test would be a table of *empirical data*, with a value for length given for each applied load. Although it is possible to work with such a table of data, it is much nicer to have the same information in the form of an equation. There are several advantages to having such an equation.

1 From an equation, it is much easier to see the functional relationship between the variables. Are they related directly or inversely? Does one vary linearly with the other, or as the square, or cube, or some other power? Is the variation exponential? A look at the equation would give the answers more readily than the tabular data.

2 An equation is a much more compact way of storing information than a table. For example, if you wished to store the results of the wire experiment in a computer for further processing, you could enter the single equation rather than the entire table.

3   If you required intermediate values between the ones measured in the test you would have to interpolate between the original data points by one of the methods of the previous chapter.   With an equation, however, interpolation becomes unnecessary.

4   Having the equation simplifies later mathematical operations such as finding roots, differentiation, and integration.

The process of fitting an equation to a set of empirical data is called *curve fitting*; the resulting equation is called an *empirical formula*, to distinguish it from equations derived from basic physical or mathematical laws.   Although not as general or as elegant as those equations derived from basic principles, and often being valid only for the particular set of conditions under which the experiment was made, these empirical formulas are extremely valuable and are widely used in every branch of technology.

9-1   HOW GOOD IS YOUR DATA?

Suppose we have a table of data and would like to find an equation that will "fit" the data.   Before we can proceed, we must ask a few questions.   First, do we know the form of the equation which will best fit our data?   Very often we know from the nature of the experiment or from the quantities being measured that the data should plot as a straight line, or a parabola, or a logarithmic curve, etc.   If we know the type of equation needed, it just remains to determine the constants in that equation to produce the best fit, after we decide what we mean by "best fit."   Techniques for fitting an equation of known form to a given set of data are given in Secs. 9-4 and 9-5.

If we have no idea of the form of the equation to try to fit to our data, our job becomes much harder.   Now we must somehow choose such a function from the countless possible choices, with no assurance that we will find one that fits.   If none of the common functions works it may be necessary to invent a new one.   We shall refer to an equation which we will attempt to fit to a set of data as a *trial formula*. In a particular problem we may have several trial formulas from which we pick the best-fitting one.

But before going to the trouble of finding or inventing an equation, let us ask whether the data are of high enough quality to justify the labor.   Do we have enough data points, and are they in close enough agreement, to clearly define the shape of the curve?   Suppose we were given the data points of Fig. 9-1a and knew nothing about the behavior of the quantities they represent.   Shall we fit them with a straight line (Fig. 9-1b)?   Or shall we fit them with a curve, and if so, which one (Fig. 9-1c and

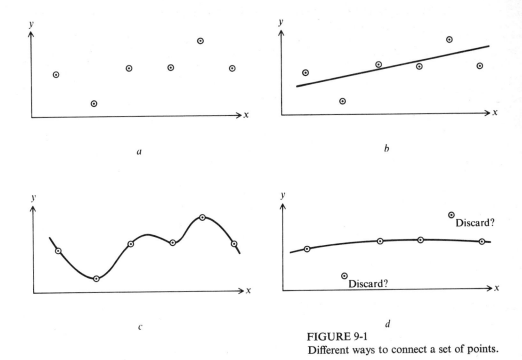

FIGURE 9-1
Different ways to connect a set of points.

d)? Shall we ignore certain points, assuming they are in error, or pass our curve through every point? If we cannot manually draw a curve through our data points with a reasonable degree of certainty, we cannot expect the computer to do any better. The data in this example are too sparse to warrant the search for an equation, whereas the data of Fig. 9-2 leave little doubt about the shape of the function.

Methods for choosing a trial formula are given in Secs. 9-6 and 9-7. It is suggested that curve fitting not be attempted with poor data.

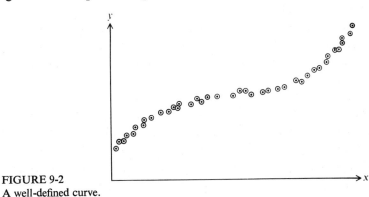

FIGURE 9-2
A well-defined curve.

9-2 MEASURING THE QUALITY OF FIT

Suppose we wish to fit our set of data points with some trial formula, $y = f(x)$, that we have obtained by some means or other. The equation will in general not fit the data exactly, and it is useful to have some measure of the quality of the fit. We may, for instance, try a second equation. And how do we determine which is the better fit?

Let us define a *residual* (also called a *difference, deviation*, or *error*) as the vertical distance between a data point and the approximating curve, as shown in Fig. 9-3. Thus,

$$r_1 = y_1 - f(x_1) \qquad (9\text{-}1)$$

The residuals can be positive, negative, or zero, and one obvious measure of the quality of fit is the algebraic sum of the residuals, which we will denote by $\sum r$. In fact, we may shift the approximating curve in such a way as to make the sum of the residuals equal to zero. This is done by varying the constants in the chosen equation in a certain way and is called the *method of averages*.

The disadvantage of using the sum of the residuals as a measure of quality of fit is evident from Fig. 9-4. If we fit the straight line AB to the three data points, the sum of the residuals is

$$\sum r = 1 + 1 - 2 = 0$$

but we would never convince anyone that this is a good fit. The dashed curve is obviously better.

Instead, we might try to minimize the sum of the *absolute values* of the residuals. In the previous example we would get

$$\sum |r| = 1 + 1 + 2 = 4$$

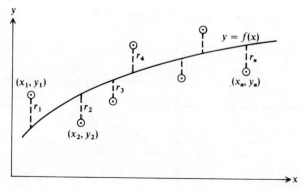

FIGURE 9-3
Definition of a residual.

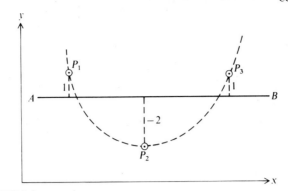

FIGURE 9-4
Sum of the residuals is zero.

which would then indicate a poor fit.   This method presents other difficulties, however, for if we look at Fig. 9-5 we see that, for line CD, $\sum |r| = 4$.   Thus if we used $\sum |r|$ as a measure of quality of fit it would indicate that line CD is as good a fit as line EF, but our own eyes tell us differently.

Suppose we *square* the residuals before adding them.   We would get, for line CD,

$$\sum r^2 = 2^2 + 2^2 = 8$$

and for line EF

$$\sum r^2 = 1^2 + 1^2 + 1^2 + 1^2 = 4$$

We see that $\sum r^2$ for the better-fitting line is lower.   In general, the sum of the squares of the residuals is found to be a reliable index of quality of fit.   The constants in our trial formula may be adjusted so as to *minimize* $\sum r^2$.   This is the *method of least squares* and is discussed in detail in Sec. 9-4.

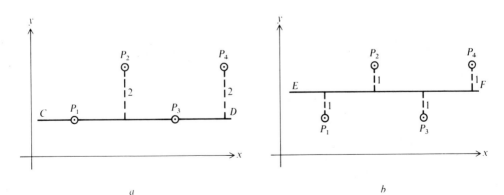

FIGURE 9-5
Adding the squares of the residuals shows that line *EF* is the better fit.

Thus our best measure of quality of fit will be the sum of the squares of the residuals, $\sum r^2$. It is also useful to know the (1) residual at each data point, (2) sum of the absolute values of the residuals (which we will call the "sum of the residuals," with it being understood that absolute values are being taken), $\sum |r|$, and (3) *average residual*, $\sum |r|/n$, where n is the number of data points. Another useful set of figures is the percent deviation between the data point and the fitting curve, at each point. These will be computed by dividing the absolute value of the residual by the ordinate of the data point and multiplying by 100.

$$Percent\ deviation = \frac{|r|}{y}\ 100$$

With these figures, we can quickly tell whether our trial formula matches our data points within their range of uncertainty.

9-3 A FORMULA EVALUATION PROGRAM

In the sections to follow, we will be writing trial formulas intended to fit a given set of data points. It will be very useful to have a program into which we can enter this formula and the given data points and have the computer tell us how good the fit is by printing the residuals, their sum, the average residual, the sum of their squares, and the percent deviation. The program is not difficult, and the student should write it in the particular language he is using, using the flowchart of Fig. 9-6 as a guide. It should be carefully saved for future use.

Let us evaluate a given formula by hand computation; the numbers in this example can be used to check our evaluation program when it is written.

EXAMPLE 9-1 The equation $y = 3x^2$ has been proposed as a possible fit for the data given in the first two columns of Table 9-1. Determine the quality of fit by computing the residuals, their sum, the sum of the squares of the residuals, the average residual, and the percent deviation.

SOLUTION We first compute $f(x) = 3x^2$ for each value of x, and enter these figures in the third column. Computing the residuals, we get

$$r_1 = y_1 - f(x_1) = 0.5 - 0 = 0.5$$
$$r_2 = y_2 - f(x_2) = 2.88 - 3 = -0.12$$
$$r_3 = y_3 - f(x_3) = 13.0 - 12 = 1$$

. . . .

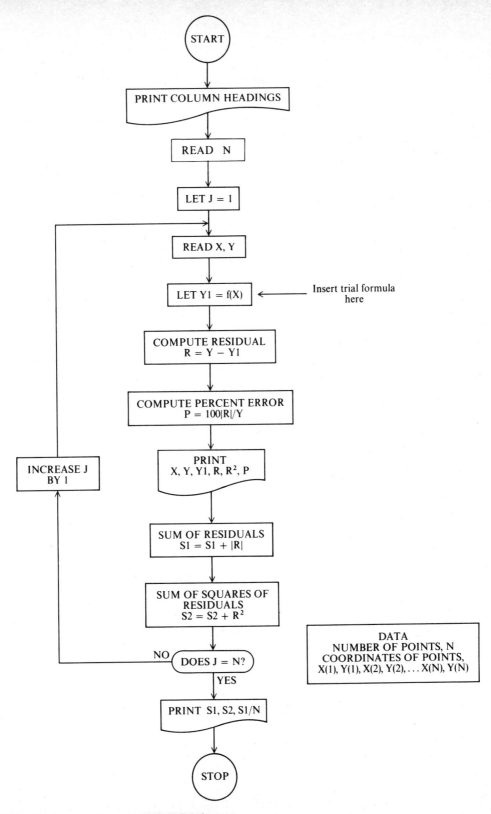

FIGURE 9-6
Flowchart for a formula evaluation program.

Table 9-1

x	y	$f(x)$	Residual	Residual squared	Percent deviation
0	0.5	0	0.5	0.25	100
1	2.88	3	−0.12	0.014	4.2
2	13.0	12	1	1	7.7
3	27.4	27	0.4	0.16	1.5
4	47.8	48	−0.2	0.04	4.2
5	75.2	75	0.2	0.04	2.7
6	106.9	108	−1.1	1.21	1.0
7	147.1	147	0.1	0.01	0.07
8	191.2	192	−0.8	0.64	0.4
9	242.5	243	−0.5	0.25	0.2
10	301.1	300	1.1	1.21	0.3

and so on for each data point. The values of the residuals are in the fourth column, the squares of the residuals are in the fifth column, and the percent deviation is in the sixth column.

Summing the fifth column we get

$$\textit{Sum of squares of residuals} = 4.824$$

Summing the absolute values of the figures in the fourth column, we get

$$\textit{Sum of absolute values of residuals} = 6.02$$

and dividing this figure by the number of data points, we get

$$\textit{Average residual} = \frac{6.02}{11} = 0.547 \qquad\qquad ////$$

With these figures we have a good measure of how well our trial formula fits the original data. To decide whether the formula is "good enough," we must take into account the uncertainty in the data and the particular application for which the formula is intended.

EXERCISE 9-1

1 Write a formula evaluation program as described in Sec. 9-3. Check your program by using it to repeat the computation of Example 9-1.

2 Compute the residuals, etc., for the following formula and set of data:

$$y = 8.82(10^{-7})x^{3.09}$$

x	273	283	288	293	313	333	353	373
y	29.4	33.3	35.2	37.2	45.8	55.2	65.6	77.3

3 The two equations

$$y = 70.63 + 0.291x$$

and

$$y = 70.59 + 0.293x$$

have been proposed as a fit for the data

x	19	25	30	36	40	45	50
y	76.30	77.80	79.75	80.80	82.35	83.90	85.10

Determine which equation is the better fit.

9-4 FITTING A STRAIGHT LINE TO A SET OF DATA POINTS

This section will describe how to find the equation of the straight line that is the best fit for a particular set of data. Many sets of data can be fitted with a straight line. More importantly, a large number of other equations can be rewritten in the form of a straight line and can thus be fitted to data by means of the simple methods in this section. These *linear transformations* will be described in Sec. 9-5.

GRAPHICAL METHOD In this method, also called the *method of selected points*, we simply plot the data points, draw a straight line with a straightedge, and read the slope m and y-intercept b of the drawn line. The equation of the line, in slope-intercept form, is then

$$y = mx + b$$

If the line does not intercept the y axis and is too far from it to be extended, we read the slope of the line and the coordinates (x_1, y_1) of any point *on the line*. Then, using the point-slope form of the equation of a straight line, we get

$$m = \frac{y - y_1}{x - x_1}$$

which can be rearranged into slope-intercept form as

$$y = mx + (y_1 - mx_1)$$

The graphical method is fast and simple and can give quite good results if done carefully. It is useful for an approximate check on the more accurate method of least squares.

METHOD OF LEAST SQUARES   A second means of determining the constants m and b is the *method of least squares* which requires that we *minimize* the sum of the squares of the residuals, or

$$\sum r^2 = \sum [\, y - f(x)]^2 = minimum$$

From calculus, we know that in order to find the minimum of some function we must take the first derivative and set it equal to zero.   Taking (partial) derivatives with respect to b,

$$\frac{\partial}{\partial b} \sum r^2 = \frac{\partial}{\partial b} \sum [y - f(x)]^2$$

$$= \frac{\partial}{\partial b} \left[\sum (y - mx - b)^2 \right]$$

$$= \frac{\partial}{\partial b} [(y_1 - mx_1 - b)^2 + (y_2 - mx_2 - b)^2 + \cdots + (y_n - mx_n - b)^2]$$

$$= -2(y_1 - mx_1 - b) - 2(y_2 - mx_2 - b) - \cdots - 2(y_n - mx_n - b)$$

$$= -2 \sum y + 2m \sum x + 2nb$$

where n is the number of data points.   Setting this equal to zero, we get

$$m \sum x + nb = \sum y \tag{9-2}$$

Now taking partial derivatives with respect to m,

$$\frac{\partial}{\partial m} \left[\sum (y - mx - b)^2 \right]$$

$$= -2(y_1 - mx_1 - b)x_1 - 2(y_2 - mx_2 - b)x_2 - \cdots - 2(y_n - mx_n - b)x_n$$

$$= -2 \sum xy + 2m \sum x^2 + 2b \sum x$$

and setting this also to zero, we get

$$m \sum x^2 + b \sum x = \sum xy \tag{9-3}$$

Equations (9-2) and (9-3) are called the normal equations, and their simultaneous solution by means of Eqs. (7-2) and (7-3) will yield the values of m and b:

$$Slope \; m = \frac{n \sum xy - \sum x \sum y}{n \sum x^2 - \left(\sum x \right)^2}$$

$$y\text{-}intercept \; b = \frac{\sum x^2 \sum y - \sum x \sum xy}{n \sum x^2 - \left(\sum x \right)^2} \tag{9-4}$$

The use of these equations will be made clearer by a numerical example.

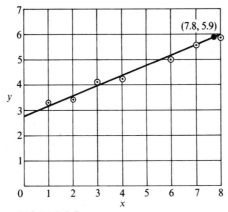

FIGURE 9-7
Plot of the data for Example 9-2.

EXAMPLE 9-2 Find the equation of the straight line that will best fit the eight data points given in the first two columns of Table 9-2, by the method of least squares. Check the results graphically.

SOLUTION

GRAPHICAL We carefully plot the points in Fig. 9-7, draw a straight line among the points, and extend the line to the y axis where we read

$$y \; intercept = 2.8$$

Selecting another point (7.8,5.9) which is on the line and far from b, we get

$$Slope = \frac{5.9 - 2.8}{7.8} = 0.40$$

Thus the equation of the line is roughly

$$y = 0.40x + 2.8$$

We shall use this equation as an approximate check on the numbers obtained by the method of least squares.

METHOD OF LEAST SQUARES In the third column of Table 9-2 we tabulate the squares of the abscissas given in column 1, and in the fourth column we list the products of x and y. The sums are given below each column.

Table 9-2

x	y	x^2	xy
1	3.249	1	3.249
2	3.522	4	7.044
3	4.026	9	12.08
4	4.332	16	17.33
5	4.900	25	24.50
6	5.121	36	30.73
7	5.601	49	39.21
8	5.898	64	47.18

$\Sigma x = 36$ $\Sigma y = 36.649$ $\Sigma x^2 = 204$ $\Sigma xy = 181.3$

Substituting these sums into Eq. (9-4), and letting $n = 8$,

$$Slope = \frac{8(181.3) - 36(36.65)}{8(204) - (36)^2} = 0.3904$$

and

$$y\ intercept = \frac{204(36.65) - 36(181.3)}{8(204) - (36)^2} = 2.824$$

which agree well with our graphically obtained values. The equation of our best-fitting line is then

$$y = 0.3904x + 2.824$$

Residuals were computed for this equation and, for comparison, for the graphically obtained equation, with the following results:

	Graphical	Least squares
Sum of squares of residuals	0.040	0.034
Sum of residuals	0.503	0.464
Average residual	0.063	0.058

////

A flowchart for computing the slope and y intercept of a straight line by the method of least squares is given in Fig. 9-8. Using the flowchart as a guide, you

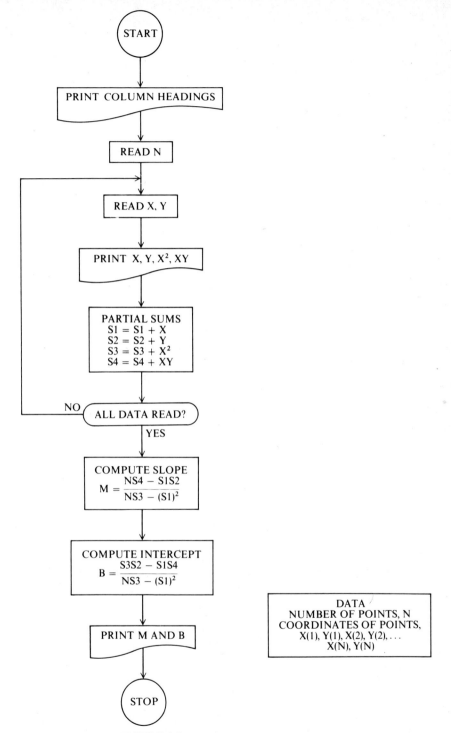

FIGURE 9-8
Flowchart for computing a least-squares line.

should write the program and debug it by using the figures in the previous example. Save the program, for it will be needed again later in this chapter.

EXERCISE 9-2

1 Plot the following lines on rectangular graph paper and determine the slope and *y* intercept.

(*a*)

x	y
−8	−6.238
−6.66	−3.709
−5.33	−0.712
−4.	1.887
−2.66	4.628
−1.33	7.416
0	10.2
1.33	12.93
2.66	15.70
4.	18.47
5.33	21.32
6.66	23.94
8.	26.70
9.33	29.61
10.6	32.35
12.	35.22

(*b*)

x	y
−20	82.29
−18.5	73.15
−17.0	68.11
−15.6	59.31
−14.1	53.65
−12.6	45.90
−11.2	38.69
−9.73	32.62
−8.26	24.69
−6.8	18.03
−5.33	11.31
−3.86	3.981
−2.4	−2.968
−0.93	−9.986
0.53	−16.92
2.	−23.86

(c)

x	y
−11.0	−65.30
−9.33	−56.78
−7.66	−47.26
−6.00	−37.21
−4.33	−27.90
−2.66	−18.39
−1.00	−9.277
0.66	0.081
2.33	9.404
4.00	18.93
5.66	27.86
7.33	37.78
9.00	46.64
10.6	56.69
12.3	64.74
14.0	75.84

(d)

x	y
5	6.882
11.2	−7.623
17.4	−22.45
23.6	−36.09
29.8	−51.13
36.0	−64.24
42.2	−79.44
48.4	−94.04
54.6	−107.8
60.8	−122.8
67.0	−138.6
73.2	−151.0
79.4	−165.3
85.6	−177.6
91.8	−193.9
98.	−208.9

2   Write a program to compute the slope and y intercept, using the method of least squares. Use your program to fit straight lines to the sets of data points in Prob. 1.

9-5   LINEAR TRANSFORMATIONS

In the preceding section we saw how to fit the best straight line to any set of data points. However, many sets of data cannot be well fitted with a straight line but require a more complex formula.   You may search through a glove store and find the pair that best fits your feet, but they will be a poor substitute for shoes.

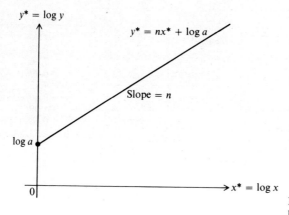

FIGURE 9-9
Plot of transformed variables.

Fortunately, of the many common formulas that may be chosen for curve fitting, many can be put into linear form by a suitable change of variables.   This enables us to fit them as we would fit any straight line.   For instance, the power function

$$y = ax^n$$

can be linearized by taking logarithms of both sides:

$$\log y = \log (ax^n) = \log a + \log x^n$$
$$\log y = n \log x + \log a$$

This will be easily recognized as the equation of a straight line in slope-intercept form if we make the following change of variables: Let

$$x^* = \log x$$
$$y^* = \log y \qquad (9\text{-}5)$$

Our equation then becomes

$$y^* = nx^* + \log a \qquad (9\text{-}6)$$

If we now plot x^* versus y^* instead of our original x versus y, our plot will be a straight line of slope n which intersects the y^* axis at $\log a$, as in Fig. 9-9.   The approximate values of the slope and y intercept can be read right off the plot, or they can be determined more accurately by the method of least squares by using Eqs. (9-4).

EXAMPLE 9-3   We have the following set of data points

x	1	2	3	4	5	6	7	8	9	10
y	1.81	2.73	3.44	4.15	4.74	5.27	5.80	6.29	6.73	7.17

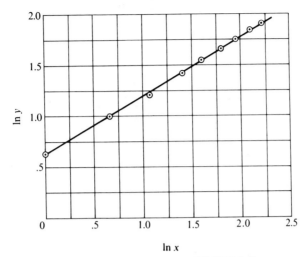

FIGURE 9-10
Plot of transformed variables in Example 9-3.

which we want to fit with an equation of the form $y = ax^n$. Verify that this is a good choice of equation by making a linear transformation and plotting the transformed variables. Then determine the constants a and n in the equation.

SOLUTION We have seen that the linear transformation for the given power function requires the plot of log x versus log y. We thus take logarithms of our data; the values are given in Table 9-3. Since logs to any base would work equally well, we have here taken natural logarithms.

Table 9-3

x	y	$x^* = \ln x$	$y^* = \ln y$
1	1.81	0	0.59333
2	2.73	0.69315	1.00430
3	3.44	1.09861	1.23547
4	4.15	1.38629	1.42311
5	4.74	1.60944	1.55604
6	5.27	1.79176	1.66203
7	5.80	1.94591	1.75786
8	6.29	2.07944	1.83896
9	6.73	2.19722	1.90658
10	7.17	2.30259	1.96991

Table 9-4

x^*	y^*	$(x^*)^2$	$x^* y^*$
0	0.59333	0	0
0.69315	1.00430	0.48046	0.69613
1.09861	1.23547	1.20694	1.35730
1.38629	1.42311	1.92180	1.97284
1.60944	1.55604	2.59030	2.50435
1.79176	1.66203	3.21040	2.97796
1.94591	1.75786	3.78657	3.42064
2.07944	1.83896	4.32407	3.82401
2.19722	1.90658	4.82778	4.18918
2.30259	1.96991	5.30192	4.53590
Sums 15.1044	14.9476	27.6502	25.4783

Plotting x^* versus y^* we get a reasonably straight line with a slope of approximately 0.6 and a y intercept of approximately 0.59, as seen in Fig. 9-10.   Since the plot of our transformed variables is nearly straight, we conclude that the trial formula $y = ax^n$ is a good choice.   We now proceed to find closer values for the slope and the y intercept by the method of least squares.

From Eqs. (9-4) we get

Slope: $\qquad\qquad m = 0.599855 = n$

y intercept: $\qquad\qquad b = 0.588713 = \ln a$

Hence,

$$a = \text{antilog } (0.588713) = 1.802$$

Thus our x fitting equation $y = ax^n$ becomes

$$y = 1.80x^{0.600}$$

after rounding a and n to the three significant figures of the original data.   ////

We have thus far done a linear transformation only of the power function. Most of the simple functions that are useful for curve fitting, however, can be similarly transformed.   A tabulation of these functions, as well as their transformations, is given in Sec. 9-7 (Fig. 9-12).   To fit an equation which cannot be transformed, techniques which are beyond the scope of this book must be used.

EXERCISE 9-3

1 Given below are several sets of data, and for each set is given a trial formula. Perform the linear transformation as shown in Fig. 9-12 and plot the transformed variables. From your plot, read the approximate values for the slope and y intercept and use these values to determine the constants in the trial formula.

(*a*)

$y = ax^n$

x	y
1	11.8450
2	59.1640
3	153.340
4	290.887
5	496.151
6	754.976
7	1097.71
8	1504.27
9	1902.81
10	2433.38

Ans. $a = 11.88, n = 2.32$

(*b*)

$y = ax^n + c$

x	y
1	22.9293
2	78.593
3	323.753
4	1002.44
5	2482.95
6	5196.11
7	9971.47
8	17225.7
9	28149.0
10	42207.3

Ans. $a = 3.33, n = 4.11, c = 19.6$

(*c*)

$y = ae^{nx}$

x	y
1	0.019243
2	0.032379
3	0.051047
4	0.080445
5	0.131351
6	0.215894
7	0.332698
8	0.547334
9	0.879082
10	1.421400

Ans. $a = 0.012, n = 0.474$

(d)

$$y = ae^{nx} + c$$

x	y
1	8.26727
2	9.3230
3	10.0338
4	11.1294
5	11.8608
6	13.2089
7	14.7906
8	15.6869
9	17.2554
10	19.2407

Ans. $a = 6.31, n = 0.398, c = 1.10$

(e)

$$y = \frac{a}{x} + c$$

x	y
1	0.020157
2	0.016686
3	0.016306
4	0.015646
5	0.015573
6	0.014832
7	0.015296
8	0.015076
9	0.014506
10	0.014614

Ans. $a = 0.00592, c = 0.0141$

(f)

$$y = \frac{1}{ax + c}$$

x	y
1	0.167414
2	0.014477
3	0.078724
4	0.060507
5	0.051489
6	0.042363
7	0.036821
8	0.032755
9	0.029711
10	0.072416

Ans. $a = 2.21, c = 7.15$

(g)

x	y
1	0.009713
2	0.016035
3	0.020825
4	0.024278
5	0.027005
6	0.028935
7	0.030929
8	0.032559
9	0.034469
10	0.035699

$$y = \frac{x}{ax + c}$$

Ans. $a = 19.8, c = 85.4$

2   Use the method of least squares to determine the constants in the trial formulas of Prob. 1 above.

9-6   SELECTING THE TRIAL FORMULA

We have seen that once we have decided which formula to use to fit our set of data points, it is not too difficult to determine the coefficients in that formula, as long as the formula is one that can be written in linear form.   But how does one decide which equation to try?   And after fitting a formula to a set of data, how does one know that some other equation might not give a better fit?

There is no general method for choosing a trial formula and no guarantee that a simple equation that will give a satisfactory fit to the data even exists.   We know that we could always write a Lagrange interpolation polynomial to pass exactly through all our data points, no matter how many we have.   However, the equation we would get would be too complex to be useful if we have more than just a few points.   Although we can give no surefire technique to follow, our bag of tricks is not empty either.   A plot of the data will often give clues to the form of the equation.   Its usefulness can then be verified by replotting the data after a suitable transformation, as was described in the previous section.   Often the nature of the experiment from which the data were obtained will suggest the form of the trial formula.   Unless the experiment is something completely brand new (which is unlikely) it may be possible to find the *form* of the equation in the literature, but the coefficients must be found for the particular set of conditions.

9-7 DEDUCING FORM FROM A PLOT OF THE DATA

The first step is to make a careful plot of the data on rectangular coordinate paper. Choose the scales so that the data points are well spread over the entire sheet. Plot each data point as a dot; indicate the uncertainty in the data by surrounding each point by a rectangle or an ellipse whose horizontal dimension is equal to the uncertainty in the abscissa and whose vertical dimension is equal to the uncertainty in the ordinate. Draw a smooth, continuous curve that passes through as many of the rectangles as possible. Have the others evenly distributed above and below the line. Use the simplest curve that will connect the data, trying the straight line first. If your knowledge of the experiment tells you the dependent variable must be zero when the independent variable is zero, make your curve pass through the origin, even if it means missing some other data points. Do the data points clearly define a curve? If they are too sparse or irregular it may be unwise to attempt curve fitting at all until better data are available. Are there one or two points lying far from the main trend? You will have to decide whether the curve should pass through or near these points or whether they should be considered *blunders* and discarded.

Examine the plot and note any features such as shown in Fig. 9-11. Does the curve cross the x or y axes, or can you extend or *extrapolate* the curve so that it does? Mark these *intercepts* on the graph. Look for *axes of symmetry, asymptotes, maximum and minimum points*, and *points of inflection*. Note whether the curve is *periodic*. With these clues and some background in algebra and analytic geometry, it is often possible to choose an appropriate trial formula. To refresh your memory of the shapes of various functions, refer to Fig. 9-12. This chart shows the plots of many common functions, as well as the special features mentioned above. When trying to match your graph to one of the curves in Fig. 9-12, realize that your data may cover only a small portion of one of these plots so that many of the special features may be missing. Also remember that you are not trying to deduce the basic relationship between the variables in your problem (although it would be very nice if you did) but are merely trying to find some function that will approximate your data over its limited range.

Should you find one or more functions that seem promising, you can check their suitability and pick out the best one by making a linear transformation, as described in Sec. 9-5. Using the original coordinates of your data points, x and y, you may compute the new coordinates x^* and y^* so that a plot of x^* versus y^* will be a straight line if your chosen trial function is a good approximation to the data. The proper linear transformations are also given in Fig. 9-12 for each of the tabulated functions. The computation of the new variables x^* and y^* can be accomplished with a simple program and then plotted to see if a (more or less) straight line results. If several trial functions are being considered, they can all be plotted and the one that gives the straighter line can be chosen.   The straight line, $y^* = mx^* + b$, can now be fitted to the original data in order to determine the constants m and b, by means of the method of least squares or simply by reading m and b from the plot, if maximum accuracy is not needed.   The constants in the trial formula can then be found by reversing the transformation, and, finally, the quality of fit found by computing residuals.   Do. not be discouraged if this seems confusing, for we have just brought together several separate computations, each one by itself quite involved.   Let us try to clear up the procedure by means of a numerical example.

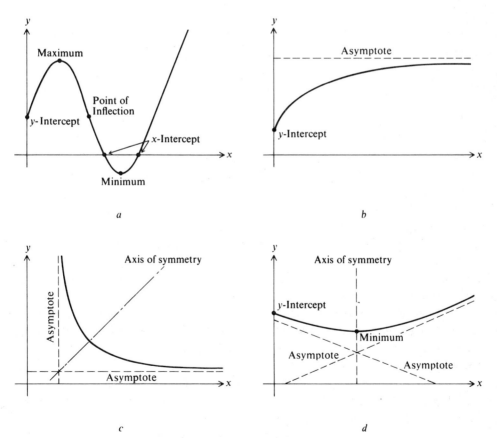

FIGURE 9-11
Special features to look for after plotting your data.

EXAMPLE 9-4   The compressive strength of concrete continues to increase for a long time after pouring, provided that it is kept moist.   Tests on such a moist cured batch of concrete yielded the following table of data, where the compressive strength is expressed as a percentage of the strength at 28 days.

Age, days	Compressive strength, percent[1]
1	14
3	33
7	59
14	80
21	91
28	100
40	102
50	119
100	133
150	140
200	146
250	149
300	152
350	153

[1] The uncertainty in the age is $\pm\frac{1}{2}$ day and the uncertainty in the strength measurement is ± 10 percent.

Fit an equation to this set of data and determine the quality of the fit.

SOLUTION   Plotting the data on rectangular graph paper, we obtain Fig. 9-13. Since the uncertainty in the abscissa is so small, our "rectangle" of uncertainty appears as a vertical straight line, extending above and below the data point by an amount equal to 10 percent of the value of the ordinate.   We draw a smooth curve through the points and, knowing that the compressive strength of concrete when first poured must be zero, we pass the curve through the origin.   We also know that if the cure were to continue for an infinite length of time we would not expect the strength to become infinite, but to approach some upper limit.   We therefore suspect the curve will approach some asymptote from below, and the shape of our plotted curve bears out this suspicion.   We pencil in an asymptote at a strength of 154 percent as an approximate location.   We also choose to discard the point (40,102) as a probable blunder.

Looking through the chart of functions (Fig. 9-12) we find two likely candidates: the exponential function $y = c(1 - e^{-nx})$, No. 6, and the hyperbolic function $y = x/(ax + c)$, No. 12.

	EQUATION and GRAPH	For Linear Transformation $y^* = mx^* + b$, let				COMMENTS
		Ordinate $y^* =$	Slope $m =$	Absicca $x^* =$	y-intercept $b =$	
Parabolic	1. $y = ax^n$ $n > 0$ 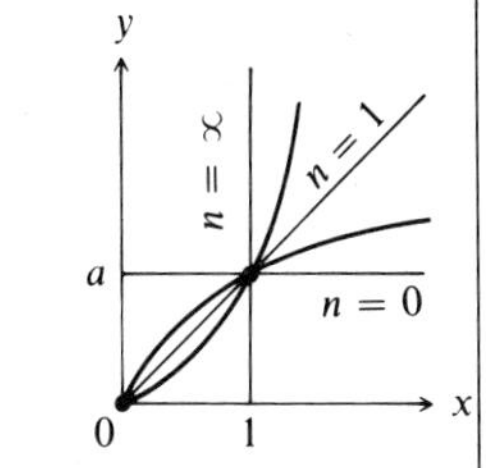	$\log y$	n	$\log x$	$\log a$	Plots linear on log-log paper
Parabolic	2. $y = ax^n + c$ $n > 0$ 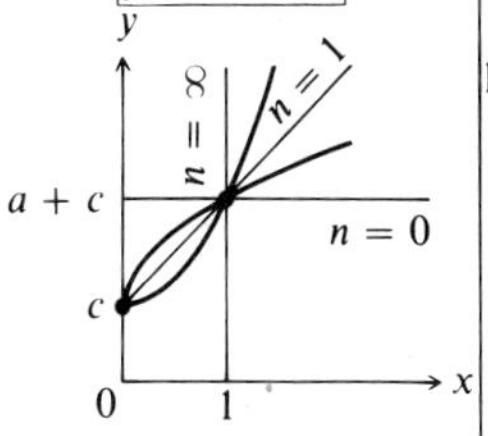	y OR $\log (y - c)$	a n	x^n $\log x$	c $\log a$	When n is known or can be guessed. Where c can be found from graph or from $$c = \frac{y_1 y_2 - y_3^2}{y_1 + y_2 - 2y_3}$$ See note 1
Parabola	3. $y = a + cx + dx^2$ 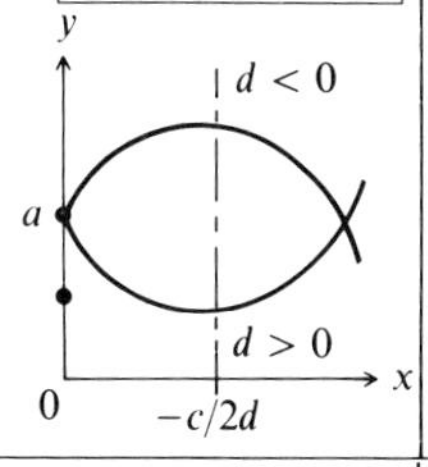	$\dfrac{y - y_k}{x - x_k}$ OR Δy	d $2dh$	x x	$c + dx_k$ $ch + dh^2$	Where (x_k, y_k) is any point on the curve. For evenly spaced x with interval h, and Δy's are first differences of the ordinates.
Exponential	4. $y = ae^{nx}$ 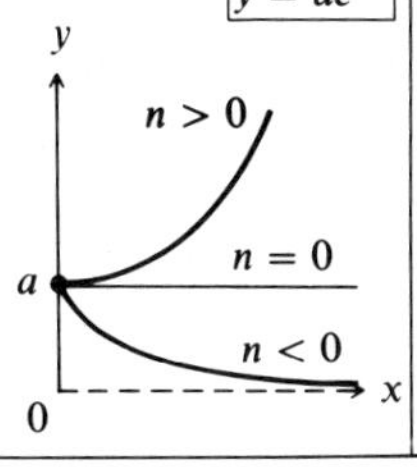	$\log y$	$n \log e$	x	$\log a$	Plots linear on semilog paper. Same as $y = ad^x$ where $d = e^n$. Use natural logs so that $\log e = 1$

FIGURE 9-12
Chart of common algebraic functions and their linear transformations.

Exponential	5. $y = ae^{nx} + c$, $a > 0$ 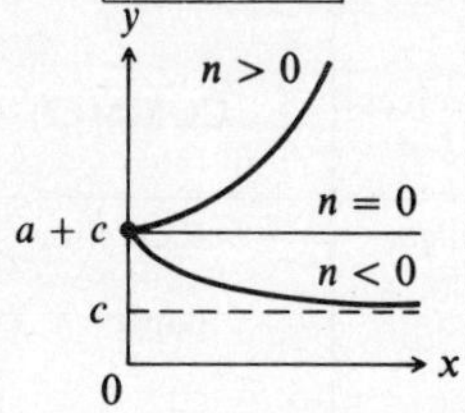	$\log(y - c)$	$n \log e$	x	$\log a$	If c can be found from graph or from $$c = \frac{y_1 y_2 - y_3^2}{y_1 + y_2 - 2y_3}$$ See note 2. Use natural logs so that $\log e = 1$
	6. $y = c(1 - e^{-nx})$, $c > 0$ 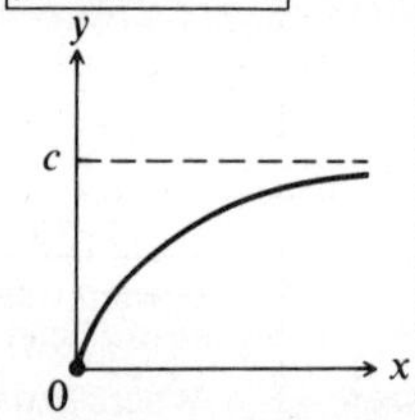	$\log(c - y)$	$-n \log e$	x	$\log c$	Find c as for equation 5. Use natural logs so that $\log e = 1$
Hyperbolic	7. $y = a/x$ 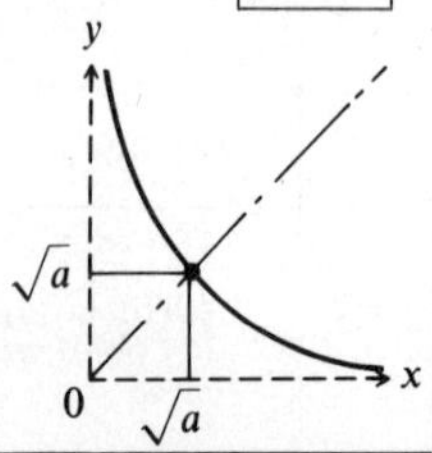	y	a	$1/x$	0	
	8. $y = a/x + c$ 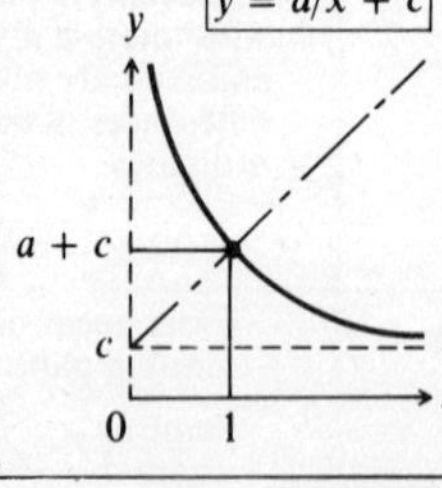	y	a	$1/x$	c	
	9. $y = a/x^n$ 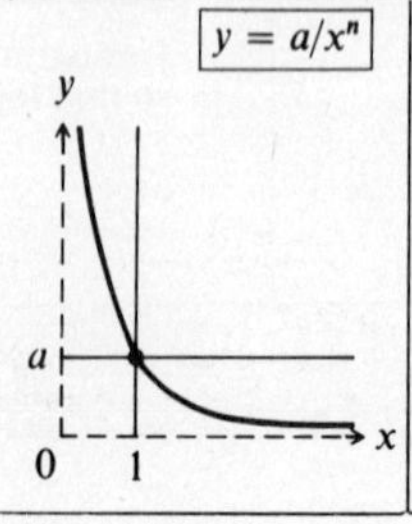	$\log y$	$-n$	$\log x$	$\log a$	Will plot linear on log–log paper

FIGURE 9.12 (*Continued*)

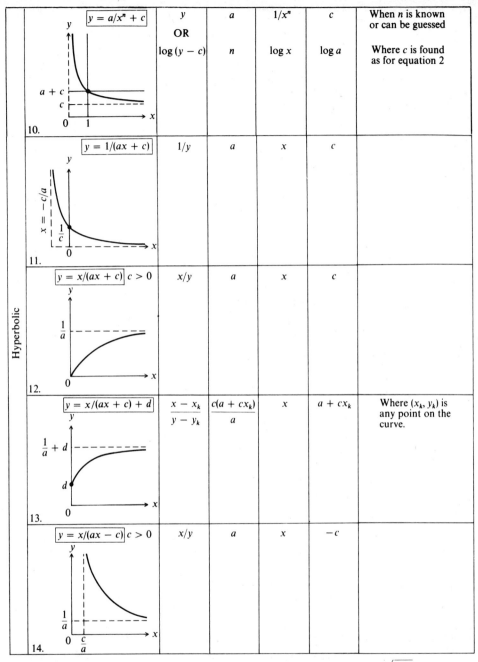

		y OR $\log(y-c)$	a n	$1/x^n$ $\log x$	c $\log a$	When n is known or can be guessed Where c is found as for equation 2
10.	$y = a/x^n + c$					
11.	$y = 1/(ax + c)$	$1/y$	a	x	c	
12.	$y = x/(ax + c)$ $c > 0$	x/y	a	x	c	
13.	$y = x/(ax + c) + d$	$\dfrac{x - x_k}{y - y_k}$	$\dfrac{c(a + cx_k)}{a}$	x	$a + cx_k$	Where (x_k, y_k) is any point on the curve.
14.	$y = x/(ax - c)$ $c > 0$	x/y	a	x	$-c$	

Note 1. Choose three points on the curve, (x_1, y_1), (x_2, y_2) and (x_3, y_3) such that $x_3 = \sqrt{x_1 x_2}$.
Note 2. Choose three points on the curve, (x_1, y_1), (x_2, y_2) and (x_3, y_3) such that $x_3 = (x_1 + x_2)/2$.
Note 3. Asymptotes are shown as dashed lines — — — — — and axes of symmetry are shown as alternating dots and dashes · — · — · —

FIGURE 9.12 (Continued)

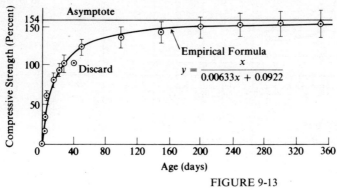

FIGURE 9-13
Compressive strength of concrete vs. age.

Looking first at the exponential function, we see that in order to linearize it we must plot x versus log $(c - y)$. We need the value of c in order to compute the new coordinates. We have an approximate value (154 percent) taken from the plot but can obtain another value from the equation

$$c = \frac{y_1 y_2 - (y_3)^2}{y_1 + y_2 - 2y_3}$$

where y_1, y_2, and y_3 are the ordinates of any three points on the curve chosen so that

$$x_3 = \frac{x_1 + x_2}{2}$$

Taking x_1 as zero and x_2 as 300, we get

$$x_3 = 150 \qquad y_1 = 0 \qquad y_2 = 152 \qquad y_3 = 140$$

and so

$$c = \frac{0 - (140)^2}{0 + 152 - 2(140)} = 153$$

which is in good agreement with our graphical value. We now compute ln $(153 - y)$ for each value of y in our table of data. These are given in Table 9-5 and are plotted with the corresponding values of x in Fig. 9-14. We see that the plot of x versus ln $(153 - y)$ does not result in a straight line.

Let us now perform the transformation for the hyperbolic function by plotting x versus x/y. These values are also listed in Table 9-5 and are plotted in Fig. 9-15.

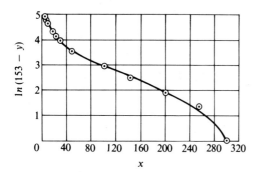

FIGURE 9-14
Plot of transformed variables x versus
ln $(153-y)$ in Example 9-4.

We see that this time our plot is quite nearly linear, and so we shall abandon the exponential form and adopt the hyperbolic function

$$y = \frac{x}{ax + c}$$

as our trial formula.   If now remains to find the values of a and c that will yield a best fit.   We shall do that by the method of least squares, but first let us find approximate values for a and c from our graphs.   For large values of x, c will become insignificant and our curve will approach the asymptote $1/a$, which we know to be approximately equal to 154.   Therefore

$$a = \tfrac{1}{154} = 0.0065 \qquad \text{approximately}$$

Table 9-5

Age x	Strength y	ln $(153 - y)$	x/y
1	14	4.93447	0.07143
3	33	4.78749	0.09091
7	57	4.56435	0.12281
14	80	4.29046	0.17500
21	91	4.12713	0.23077
28	100	3.97029	0.28000
50	119	3.52636	0.42017
100	133	2.99573	0.75188
150	140	2.56495	1.07143
200	146	1.94591	1.36986
250	149	1.38629	1.67785
300	152	0	1.97368
350	153	....	2.28758

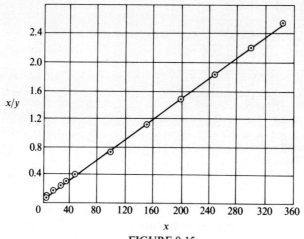

FIGURE 9-15
Plot of transformed variables x versus x/y in Example 9-4.

Also in our linearized plot, the slope of the line is equal to a

$$a = \frac{rise}{run} = \frac{2.22}{349} = 0.0063 \qquad \text{approximately}$$

and the y intercept is equal to c

$$c = 0.1 \qquad \text{approximately}$$

Applying the method of least squares, we now find the slope and y intercept of the line given by the data in the first and fourth columns of Table 9-5, and we obtain

$$Slope = a = 0.00633$$
$$y\ intercept = c = 0.0922$$

Substituting these values into our trial formula, we get

$$y = \frac{x}{0.00633x + 0.0922}$$

as our final result.

As a final step, we use our evaluation program of Sec. 9-3 and get the following:

$$Sum\ of\ residuals = 41$$
$$Average\ residual = 3.16$$
$$Sum\ of\ squares\ of\ residuals = 169$$

The percent deviation at each data point was also computed and was found to be less than the ± 10 percent uncertainty of the original data except at the first two points, where the empirical formula gave values 27 and 18 percent lower than required. The formula, then, would have to be used with caution at the lower values of x.

The formula is plotted in Fig. 9-13 to show how it fits the original data. Although a better-fitting formula can probably be found, note that it must be of a different form than $y = x/(ax + c)$, for the one we just computed is the best obtainable (in the least-squares sense) with this form.   ////

EXERCISE 9-4

1 Given below are several sets of data. Plot the data and try to match your plot with one or more of the curves in Fig. 9-11. For each of your chosen trial formulas, perform the transformation and plot the transformed variables. Pick the one producing the straightest plot and determine the slope and y intercept by the method of least squares, checking these values graphically. Use these values to determine the constants in your trial formula.

(*a*)

x	y
1	27.4443
2	55.7821
3	89.9455
4	126.496
5	170.106
6	225.678
7	275.341
8	339.273
9	397.218
10	488.647

Ans. $y = 22.1x + 2.63x^2$

(*b*)

x	y
1	2.53055
2	4.35128
3	5.60903
4	6.56518
5	7.44592
6	7.72553
7	7.92246
8	8.52895
9	8.58069
10	8.61687

Ans. $y = 8.92(1 - e^{-0.358x})$

(c)

x	y
1	28.8545
2	14.1084
3	9.31589
4	7.07553
5	5.84623
6	4.71698
7	4.13326
8	3.62351
9	3.19510
10	2.88840

Ans. $y = \dfrac{28.8}{x}$

(d)

x	y
1	10309.0
2	2221.40
3	896.393
4	478.012
5	301.872
6	194.708
7	138.969
8	104.226
9	82.6799
10	63.3439

Ans. $y = \dfrac{10{,}242}{x^{2.2}}$

(e)

x	y
1	2.11045
2	3.02312
3	3.76932
4	4.48385
5	5.15394
6	5.66389
7	6.17919
8	6.60835
9	6.95656
10	7.59805

Ans. $y = \dfrac{x}{0.066x + 0.96} + 1.15$

(f)

x	y
1	38.3106
2	20.5771
3	18.0191
4	16.6519
5	15.8487
6	15.9515
7	15.2289
8	15.5707
9	15.4870
10	14.8025

$$Ans. \quad y = \frac{x}{0.0708x - 0.0444}$$

PROJECTS

1   The elongation of a wire measured for various applied loads will yield a set of data points that will plot as a straight line, as long as the elastic limit of the material is not exceeded. Write a program that will pass a least-squares line through such a set of data and, knowing the wire length and diameter, compute the modulus of elasticity.   Run your program with data that you have collected yourself in the physics laboratory, or use the following data:

Load, lb	0	10	20	30	40	50	60	70	80	90
Elong., in.	0	0.017	0.032	0.052	0.069	0.084	0.103	0.119	0.134	0.152

Wire length = 100 in. and diameter = 0.05 in.

2   Rubber does not follow Hooke's law and thus does not have a constant modulus of elasticity as do metals, but the modulus of elasticity depends upon the hardness of the rubber. The following table gives the relationship between the durometer hardness number and the modulus of elasticity in compression, in pounds per square inch.   Fit an empirical equation to the data.

Modulus of elasticity	Durometer hardness
120	27
150	33
180	38
210	43
240	47
270	51
300	53
330	56
360	59
390	62
420	64
450	66
480	68
510	69

3 Find an empirical equation linking belt speed (feet per minute) with horsepower per inch of width for a double-ply $\frac{5}{16}$-in.-thick leather belt.

Speed	Horsepower
600	1.8
800	2.4
1000	3.1
1200	3.7
1400	4.3
1600	4.9
1800	5.4
2000	6.0
2200	6.6
2400	7.1
2600	7.7
2800	8.2
3000	8.7
3200	9.2
3400	9.7
3600	10.1
3800	10.5
4000	10.9
4200	11.3
4400	11.7
4600	12.0
4800	12.3
5000	12.5
5200	12.8
5400	12.9
5600	13.1
5800	13.2
6000	13.2

4 The distance fallen by a body under the action of gravity was measured for different elapsed times and the following data obtained:

Time, s	Distance, ft
0.1	0.284
0.2	0.772
0.3	1.45
0.4	2.36
0.5	3.42
0.6	4.73
0.7	6.19
0.8	7.86
0.9	9.76
1.0	11.8

Fit a curve to these data and thereby find the initial velocity and acceleration.   Instead of using the above data you may wish to use data collected yourself in the physics laboratory.

5 A vacuum-tube diode was found to have the following characteristics:

Voltage across diode, V	Current through diode, M/A
0	0
5	2.063
10	5.612
15	10.17
20	15.39
25	21.30
30	27.71
35	34.51
40	42.10
45	49.95
50	58.11

Fit an equation to this set of data points.   *Ans.* $I = 2 \times 10^{-4} V^{1.45}$

10

ROOTS OF EQUATIONS

10-1 LINEAR AND QUADRATIC EQUATIONS

In the solution of engineering problems, it is often necessary to find the roots of equations, which means to find the values of the independent variable when the dependent variable is zero.

If the equation is linear or quadratic, the task is easy. For a linear equation such as

$$ax + b = 0$$

the solution is simply

$$x = -\frac{b}{a}$$

and for a quadratic of the form

$$ax^2 + bx + c = 0$$

application of the quadratic formula gives

$$x = \frac{-b \pm \sqrt{b^2 - 4ac}}{2a}$$

both of these solutions being readily programmed.   For the solution of the quadratic, however, we must take the precaution of examining the discriminant ($b^2 - 4ac$). If it is negative the roots must be printed in complex form.

EXAMPLE 10-1   An instrument package is fired vertically upward into a smog layer to take atmospheric readings, which it transmits back to a receiver on the ground.   The readings are desired at 10-ft intervals from the ground.   Neglecting air resistance, find the times at which the readings should be taken, if the package is fired with an initial velocity of 100 ft/s.

SOLUTION

UNKNOWNS   Those times T (in seconds) at which the package is at 10 ft intervals of height above the ground.

DATA

1   The package height H (in feet).
2   The initial velocity V_0 equals 100 ft/s.

RELATIONSHIPS   From physics, the height of a body projected vertically upward is

$$H = V_0 T - \tfrac{1}{2}GT^2$$

where $G = 32.2$ ft/s^2.

OUTPUT FORMAT   The most suitable format would be a three-column table containing

1   The package height, in 10-ft intervals
2   The time at which the package is at height H on the ascent
3   The time at which the package is at height H on the descent

PLANNING THE SOLUTION   To compute the required times we must find the roots of the quadratic equation

$$\tfrac{1}{2}GT^2 - V_0 T + H = 0$$

for each value of H as H increases from zero in 10-ft intervals.   For each successive value of H we must see if the discriminant has become negative.   This would indicate that the maximum height of the package has been passed and we may end the computation.   We shall evaluate the roots by the quadratic formula, print the smaller root in the second column of our output table, and print the larger in the third column.

Table 10-1

Package	Time	
height	Ascent	Descent
0	0	6.2112
10	0.1017	6.1095
20	0.2069	6.0043
30	0.3161	5.8951
40	0.4097	5.7815
50	0.5484	5.6626
60	0.6729	5.5383
70	0.8041	5.4071
80	0.9432	5.2679
90	1.0919	5.1192
100	1.2526	4.9586
110	1.4286	4.7826
120	1.6253	4.5859
130	1.8525	4.3586
140	2.1314	4.0797
150	2.5330	3.6782

A flowchart for this computation is given in Fig. 10-1, and the output is tabulated in Table 10-1.

CHECKING THE RESULTS   Let us check the computed values by hand computation at one particular height, say 100 ft.   We get

$$T = \frac{V_0 - (V_0{}^2 - 2GH)^{1/2}}{G} = \frac{100 - (10{,}000 - 6440)^{1/2}}{32.2}$$

$$= \frac{40.3}{32.2} = 1.25 \quad \text{and} \quad \frac{159.7}{32.2} = 4.96$$

which agree with the computed values within slide-rule accuracy.   ////

EXERCISE 10-1

1 Write a program to find the roots of any quadratic equation.   Be sure that the program will detect a negative discriminant and print imaginary roots in complex form.   Try your program on the following equations:

(a) $x^2 - 3x + 2 = 0$ Ans. $x = 1, 2$

(b) $x^2 + 2x - 15 = 0$ Ans. $x = 3, -5$

(c) $6x^2 + 13x + 6 = 0$ Ans. $x = -\frac{2}{3}, -\frac{3}{2}$

(d) $x^2 - 2x + 2 = 0$ Ans. $x = 1 \pm j$

(e) $x^2 + x + 1 = 0$ Ans. $x = -0.5 \pm j(0.866)$

(f) $5x^2 - 3x + 1 = 0$ Ans. $x = 0.3 \pm j(0.3317)$

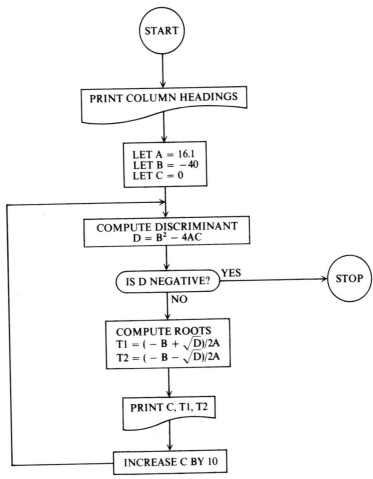

FIGURE 10-1
Flowchart for the solution of Example 10-1.

10-2 ITERATION TECHNIQUES

Unfortunately, not all equations are as simple as those given above. It may be necessary to solve an equation whose degree is higher than 2. It may be further complicated with transcendental functions, nonintegral exponents, or worse, so that direct solution may be difficult or impossible. In such cases one may resort to the method of iteration, which will usually yield results. The general iteration method was presented in Sec. 7-2 where it was applied to the solution of a set of linear equations. The remainder of this chapter will describe several methods of finding roots by iteration. All these methods follow the same general procedure:

1 Make a *first guess* at the root. Your guessing may be aided by a plot of the function or by knowledge of the physical problem. You can also use a computer "scan" of the x axis, in which you instruct the computer to print the ordinates at selected values of x over the region of interest, and note the abscissas between which the ordinate changes sign.

2 Using the first guess as a starting point, obtain a *closer approximation* to the root by means of one of the several methods described in this chapter.

3 Using the second approximation, apply the selected method again to obtain a *third approximation*.

4 Continue to obtain *successive approximations* until the desired degree of accuracy is obtained. This will be so when the difference between the latest approximation and the one immediately preceding it is smaller than the least significant digit required.

Not all the methods described here will converge under all conditions, and some will converge faster than others.

The methods described will work for polynomials as well as for transcendental functions (logarithmic, exponential, trigonometric, and hyperbolic). We will be concerned in this chapter with finding only real roots, except in the case of the quadratic equation, where imaginary roots are found as well.

10-3   METHOD OF SIMPLE ITERATION

Let us consider the equation

$$3x - 2 \log x - 5 = 0$$

The method of *simple iteration* consists of first splitting the equation into two or more terms, one of which is the variable itself, and transposing so that the variable is alone on one side of the equation. Thus, dividing through by 3 and transposing, we get

$$x = \tfrac{1}{3}(2 \log x + 5) \tag{10-1}$$

Note that we have not *solved* our equation for x because the right-hand side (RHS) still contains x. We have merely rearranged the equation into the form $x = f(x)$.

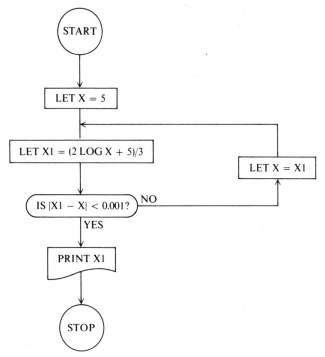

FIGURE 10-2
Flowchart for the method of simple iteration.

We now take our *first guess* at x (call it X_0), substitute it for x in the RHS of Eq. (10-1), and compute our next approximation X_1.

Thus

$$X_1 = \frac{2 \log X_0 + 5}{3}$$

The value X_1 is now substituted for X_0 in Eq. (10-1), and the next approximation X_2 is computed,

$$X_2 = \frac{2 \log X_1 + 5}{3}$$

and so on until the *change* in each successive approximation is smaller than the least significant figure required.

Figure 10-2 shows a flowchart for solving this equation, and Table 10-2 gives the results of a computation, taking 5 as the first guess. Successive values of x are also shown, as well as the difference between each successive value. Notice that there was no change in the fifth decimal place after the seventh iteration. We shall compare the rate of convergence with that of the other methods in this chapter.

The computation may not converge, and whether it converges or diverges usually depends upon how the original equation is split up. The equation of our previous example, for instance, might just as well have been written

$$\frac{3x - 5}{2} = \log x$$

or

$$x = 10^{(3x - 5)/2}$$

If we now tried to find the roots by simple iteration, our computation would *diverge*.

We may wish to insert simple statements into the program that would end the run should the computation diverge, to avoid wasting costly computer time. We might, for example, see if the differences between successive approximations were getting larger rather than smaller and halt the program in that event. We might also limit the number of iterations to some arbitrary number, say 20.

Table 10-2   SIMPLE ITERATION

Iteration	X	Difference
0	5	
		−2.86735
1	2.13265	
		−0.24670
2	1.88595	
		−0.03559
3	1.85035	
		−0.00552
4	1.84484	
		−0.00086
5	1.84397	
		−0.00014
6	1.84384	
		−0.00002
7	1.84381	
		0
8	1.84381	

10-4   METHOD OF INTERVAL HALVING

The method of *interval halving* (also called the *half-interval*, *midpoint*, and *bisection* method) requires two initial guesses, which must lie on either side of the root.   It has the advantage that it will always converge.   Referring to Fig. 10-3, the initial guesses are X_1 and X_2, from which are computed Y_1 and X_3, where

$$X_3 = \frac{X_1 + X_2}{2}$$

The ordinate Y_3 is then computed and its sign compared with the sign of Y_1.   If they have the same sign, X_1 is replaced by X_3 and the procedure repeated for the interval from X_3 to X_2.   If the signs of Y_1 and Y_3 were opposite, X_2 would be replaced with X_3 and the procedure repeated.

A program for the solution of the equation of Sec. 10-3 was written.   This program was run by taking as first approximations X_1 equal to zero and X_2 equal to 5; the results are tabulated in Table 10-3.   Notice that even after 10 iterations the third decimal place is still in doubt.

Table 10-3   HALF-INTERVAL METHOD

Iteration	x	Difference
0	0	
		3.00000
1	3	
		−1.00000
2	2	
		−0.50000
3	1.5	
		0.25000
4	1.75	
		0.12500
5	1.875	
		−0.06250
6	1.8125	
		0.03125
7	1.84375	
		−0.015634
8	1.85937	
		−0.00781
9	1.85156	
		−0.00391
10	1.84766	

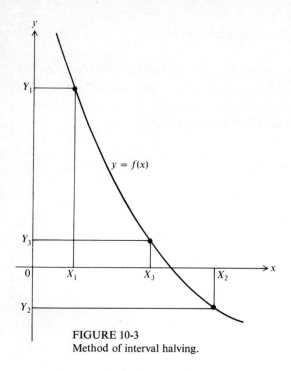

FIGURE 10-3
Method of interval halving.

A flowchart for finding roots by the method of interval halving is given in Fig. 10-4.   It is necessary to enter the two initial guesses X_1 and X_2, and the difference ε between each computed X_3 and its previous value X_0 at which you want the computation to end.   This flowchart will not cause the printing of all the intermediate values of X_3 as in Table 10-3, but only its final value.

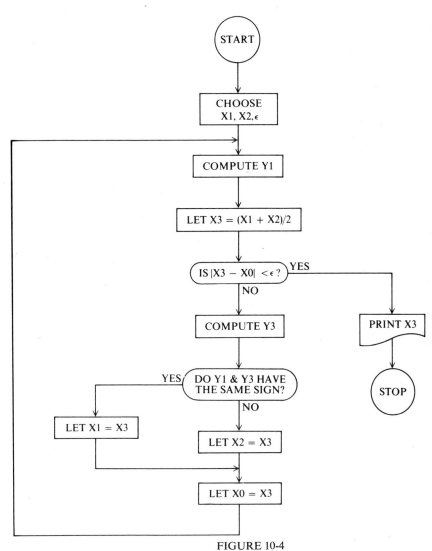

FIGURE 10-4
Flowchart for the method of interval halving.

10-5 FALSE POSITION METHOD

This method, also called *regula falsi* and the *secant method*, is similar to the method of interval halving in that it requires two initial approximations. The method will sometimes work if both initial guesses are on the same side of the root, but to guarantee convergence, they should bracket the root.

From Fig. 10-5, the two initial approximations are X_1 and X_2. The line connecting P_1 and P_2 crosses the x axis at X_3, which is taken as the next approximation to the root.

The slope of this line is

$$m = \frac{Y_2 - Y_1}{X_2 - X_1} = \frac{Y_2 - 0}{X_2 - X_3}$$

from which

$$X_3 = X_2 - Y_2 \frac{X_2 - X_1}{Y_2 - Y_1}$$

As in the interval-halving method, the sign of Y_3 is now compared with the sign of Y_1; if they are the same, X_1 is replaced by X_3, and the procedure is repeated.

Once again using the equation of Sec. 10-3, with X_1 set equal to zero and X_2 equal to 5, a program was written and the results of the computation are given in Table 10-4. Notice the much faster convergence, with the fifth decimal place being determined by the fifth iteration.

A flowchart for the false position method is given in Fig. 10-6.

Table 10-4 **FALSE POSITION METHOD**

Iteration	x	Difference
0	0	
		1.75457
1	1.75457	
		0.08260
2	1.83716	
		0.006167
3	1.84333	
		0.00045
4	1.84378	
		0.00003
5	1.84381	
		0
6	1.84381	

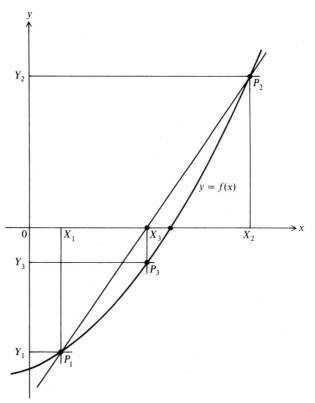

FIGURE 10-5
False position method.

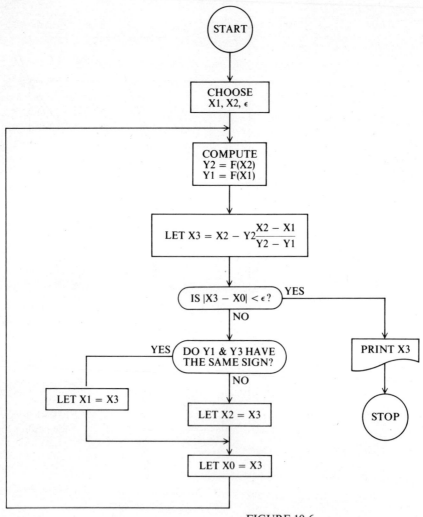

FIGURE 10-6
Flowchart for the false position method.

10-6 NEWTON-RAPHSON METHOD

The *Newton-Raphson* method of approximating roots is probably the most widely used method because of its rapid convergence and ease of programming. It requires only one first guess which, if not sufficiently close to the root, may sometimes cause divergence, or convergence on the wrong root. To use this method it is necessary to know the first derivative of the given function.

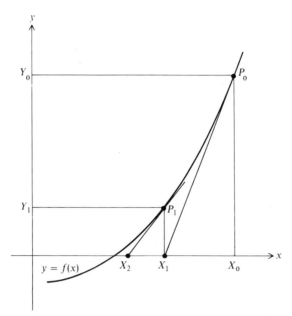

FIGURE 10-7
Newton-Raphson method.

Referring to Fig. 10-7, we make the initial guess X_0, from which we compute the ordinate Y_0. Evaluating the first derivative of our function at X_0, we obtain the slope m of the tangent to the curve at P_0, which crosses the x axis at X_1. We take X_1 as our second approximation to the root. Since

$$m = f'(X_0) = \frac{Y_0 - 0}{X_0 - X_1}$$

we get

$$X_1 = X_0 - \frac{Y_0}{m}$$

Our new approximation X_1 is now used to compute Y_1 and the slope at P_1 from which a third approximation X_2 is found, and so on until the required accuracy is achieved.

Applying the Newton-Raphson method to the equation of Sec. 10-3

$$f(x) = 3x - 2 \log x + 5$$

we take the first derivative

$$f'(x) = 3 - \frac{2}{x \ln 10} = m$$

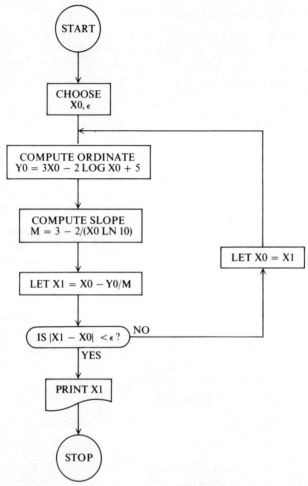

FIGURE 10-8
Flowchart for the Newton-Raphson method.

Figure 10-8 is a flowchart illustrating the use of these equations in the computation of the root. The results of the computation, again using 5 as a first guess, are given in Table 10-5. The root was found accurate to five decimal places with only three iterations, demonstrating the extremely rapid convergence of this method.

Table 10-5

Iteration	x	Difference
0	5	
		−3.04360
1	1.9564	
		−0.11201
2	1.8444	
		−0.00059
3	1.84381	
		0
4	1.84381	

EXAMPLE 10-2   A 15-ft-long cantilever beam (Fig. 10-9) has a modulus of elasticity of 9×10^6 lb/in.^{2}, a moment of inertia of 1.45 in.^{4}, and a concentrated load of 350 lb at the free end.   Find the distance from the fixed end at which the deflection will be 0.03 in.

SOLUTION

UNKNOWNS   The distance from the fixed end at which the deflection is 0.03 in.   Let X be this distance in feet.

DATA

Beam length: $L = 15$ ft
Modulus of elasticity: $E = 9 \times 10^6$ lb/in.^{2}
Moment of inertia: $I = 1.45$ in.^{4}
Load: $P = 350$ lb
Deflection: $D = 0.03$ in.

RELATIONSHIPS   From references on strength of materials, we find that the vertical deflection of a cantilever beam with a concentrated load at one end is given by the formula

$$D = \frac{PX^2}{2EI} (3L - X)$$

in which all quantities are known except X.

OUTPUT FORMAT   Our output can be a simple statement such as

THE DEFLECTION IS 0.03 IN. ____FT FROM FIXED END

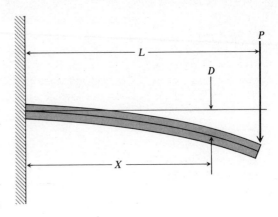

FIGURE 10-9
Cantilever beam with a concentrated
load at the end.

PLANNING THE SOLUTION   We see that our solution consists of finding a root of the
cubic equation

$$f(X) = \frac{P}{2EI}(3LX^2 - X^3) - D$$

A cubic equation will have three roots, but in this problem we are interested only in
the root lying between $X = 0$ and $X = 15$.   Let us choose to find the root by the
Newton-Raphson method, although any of the methods of this chapter would be
applicable.   Taking the first derivative, we get

$$f'(X) = \frac{P}{2EI}(6LX - 3X^2) = m$$

For a first guess for X we may take the midpoint of the beam where $X = 7.5$ ft.   We
shall halt the computation when our approximation does not change by more than
0.01 ft.   The flowchart for this computation is very similar to Fig. 10-8, and the
result obtained is 7.76 ft.

CHECKING THE RESULT   Although it is difficult to find X, it is a relatively simple
matter to check it once obtained, by substituting it back into the original equation.
Thus,

$$D = \frac{350(7.76)^2}{2(9)(10^6)(1.45)}[3(15) - 7.76] = \frac{0.0211}{26.1}(37.24) = 0.03 \text{ in.} \qquad ////$$

EXERCISE 10-2

1 The following equations have a positive real root between 0 and 10. First isolate the root by a scan of the x axis between those limits, and then find the root to four decimal places by any of the methods in this chapter.

(a) $y = \dfrac{e^x + e^{-x}}{2} - 3$ *Ans.* 1.7628

(b) $y = 2x^5 - 3x^3 + x - 1$ *Ans.* 1.2008

(c) $y = 2e^x - 7 \cos x$ *Ans.* 0.8440

(d) $y = e^x - 3 \ln x - 6$ *Ans.* 2.1088

(e) $y = 3x - 8 + 2x^x$ *Ans.* 1.4784

PROJECTS

1 A rocket is launched and follows a course given by the equation

$$y = 1.85 \sin \frac{x}{3} + e^{x/2} - 1.35x$$

where $x =$ miles traveled horizontally

$y =$ miles traveled vertically

An instrument package is released at a height of 4 mi. Find the horizontal distance from the launch site at which the package is released.   *Ans.* $x = 4.083$ mi

2 A solid steel column of circular cross section is needed to support a load of 65,000 lb. The ends of the column are not rigidly fixed but are able to rotate, and the point of application of the load may be as much as 1 in. from the axis of the column. The yield point of the steel is 50,000 lb/in.², and the desired safety factor is 3. The column is 45 in. long. Find the required diameter to safely carry the load.

The equation giving the maximum stress in a column of this type is

$$s = \frac{FP}{A}\left(L + \frac{ec}{i^2} \sec \frac{L}{2i} \sqrt{\frac{FP}{AE}}\right)$$

where $s =$ maximum stress $= 50,000$ lb/in.²

$F =$ factor of safety $= 3$

$P =$ applied load $= 65,000$ lb

$A =$ cross-section area, in.² $= \pi d^2/4$ for circular column

$e =$ eccentricity of applied load $= 1$ in.

$c =$ distance from column axis to extreme edge of cross section $= d/2$ for circular column

$i =$ radius of gyration $= d/4$ for circular column

$E =$ modulus of elasticity $= 30,000,000$ for steel

$L =$ column length $= 45$ in.   *Ans.* Diameter $= 15.04$ in.

3 The gain of a voltage divider is given by the equation

$$G = \frac{1}{\sqrt{1 + (2\pi fRC)^2}}$$

where R is the resistance (8500 Ω), C is the capacitance (1.1×10^{-6} F), and f is the frequency of the sinusoidal input voltage, in hertz.   Find the frequency at which G equals 0.85.   *Ans.* $f = 10.5$ Hz

4 A dipstick is to be used to measure, to the nearest gallon, the quantity of liquid in a hemispherical storage tank.   The tank has an inside radius of 12 in., and the stick has a cross-section area of 1.5 in.^{2} and is to contain 15 marks, indicating the liquid volume from 1 to 15 gal.   Write a program to compute the position of each gallon marking to an accuracy of 0.01 in.   Remember that the dipstick itself will displace some liquid and will affect the readings.

Check   The 10 gal mark is 9.08 in. from the end of the dipstick.

11

DIFFERENTIATION

If the functional relationship $y = f(x)$ between the variables x and y is known, it is usually easy to write the derivative $y = f'(x)$. This derivative can then be inserted into a program to enable the computation of the derivative at any particular value of x. Such derivatives, as you will recall from your calculus, are useful in all branches of technology, since they give the rate of change of the function. It can be used to find velocities, accelerations, points where the curve reaches a maximum or minimum, and many other quantities of practical interest.

11-1 ALTERNATIVE METHODS

Trouble arises, however, if the functional relationship is in the form of a table (and we do not have an equation), or if our equation is too difficult to differentiate. In such cases there are several other ways in which we may determine the derivative at any point; these alternatives are shown in Fig. 11-1.

If we know the equation and can differentiate it, we proceed to do so. If we cannot differentiate the equation, we may use any of the four remaining methods: graphical, mechanical, by definition, or by polynomial interpolation. The simplest

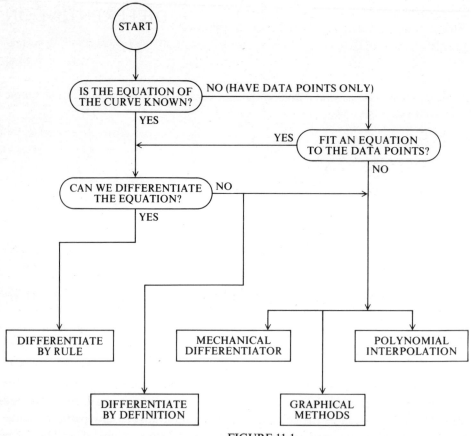

FIGURE 11-1
Choosing the method to use for finding derivatives.

of these is probably the graphical. This consists of plotting the curve, placing a first-surface mirror at right angles to the curve through the point at which the slope is needed. The mirror is adjusted so that the reflection in the mirror appears continuous with the curve on the paper, and using the surface of the mirror as a straight-edge, the normal to the curve is drawn. The derivative is then the slope of the perpendicular to this normal. In addition to this graphical method, there are mechanical devices available, in which a flexible metal strip is forced to conform to the shape of the curve, and the slope is read on the scales provided. The remaining two methods will be described later in this chapter.

If the relationship is in the form of a table of data we may choose to try to find an equation to represent the entire table of data, using the techniques of Chap. 9.

If we can write such an equation but cannot differentiate it we are no better off than before. Instead of trying to write such an empirical equation we may choose to go directly to the graphical or mechanical method, or to use the method of polynomial approximation. This technique, described in detail in Sec. 11-3, consists of writing an equation connecting a *few* of the data points by means of the Lagrange interpolating polynomial and then finding the first derivative of this polynomial (a simple matter) at the point of interest.

11-2   BY DEFINITION

Recalling the definition of a derivative, and referring to Fig. 11-2a,

$$\frac{dy}{dx} = \lim_{\Delta x \to 0} \frac{f(x + \Delta x) - f(x)}{\Delta x}$$

If, however, we take the increments Δx to either side of the point at which the derivative is sought, as in Fig. 11-2b, we may rewrite the definition

$$\frac{dy}{dx} = \lim_{\Delta x \to 0} \frac{f(x + \Delta x) - f(x - \Delta x)}{2\Delta x} \tag{11-1}$$

This equation may now be programmed, setting Δx equal to some small, finite value, and dy/dx computed. The increment Δx is now reduced in steps to smaller and smaller values, each time computing the corresponding value of dy/dx until the change in any two successive values of the derivative is less than the accuracy desired in the computation.

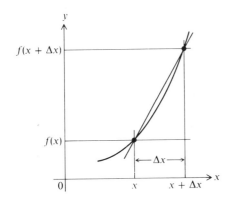
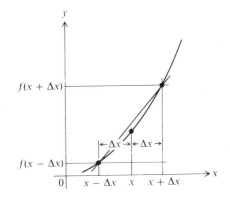

a

b

FIGURE 11-2
Derivatives by the delta method.

Notice that in the numerator of Eq. (11-1) we shall be taking the difference of two nearly equal numbers, a situation we warned against in Chap. 5 because of the drastic loss in significant figures that can occur. However, the two quantities in the numerator are computed from the given equation and therefore contain the full number of significant figures that the computer is capable of storing. Even so, we must be careful not to make Δx too small; the following example illustrates how this is done.

EXAMPLE 11-1   To illustrate the method, let us choose a function which we can differentiate, so that we may compare our results with the exact derivative. Let us find the first derivative of $y = \ln x$ for values of x from 1 to 5 in steps of 1.

We want to reduce Δx in order to obtain a greater accuracy in our computation, but not so far that the numerator in Eq. (11-1) will have fewer than, say, three significant figures. The largest value that our function $f(x)$ will assume over the given range will be about 1.6, when x equals 5. If our computer is capable of storing numbers to eight significant figures, this number will then contain seven decimal places. Thus if our numerator is to have at least three significant figures, it must not be allowed to fall below the fifth decimal place. Let us take Δx equal to 0.1 initially and reduce it by a factor of 10 for each successive computation until the numerator of Eq. (11-1) reaches a value of 0.00001. A flowchart for this program is shown in Fig. 11-3, and the results of the computation are contained in the following table.

x	Delta x	Derivative
1	0.1	1.00335
	0.01	1.00003
	0.001	1.00000
	0.0001	1.00000
2	0.1	0.500417
	0.01	0.500004
	0.001	0.500000
	0.0001	0.500000
3	0.1	0.333457
	0.01	0.333335
	0.001	0.333333
	0.0001	0.333333
4	0.1	0.250052
	0.01	0.250001
	0.001	0.250000
	0.0001	0.250000
5	0.1	0.200027
	0.01	0.200000
	0.001	0.200000
	0.0001	0.200000

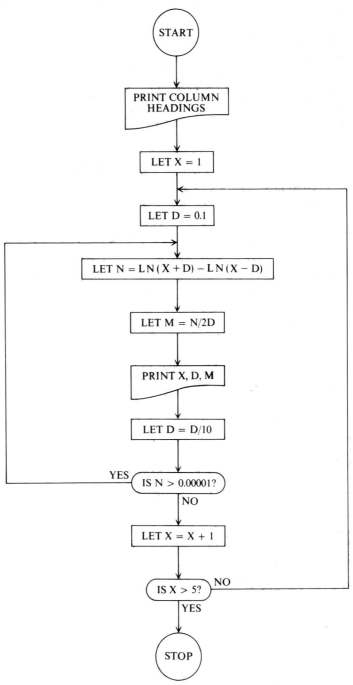

FIGURE 11-3
Flowchart for Example 11-1.

The exact values of the derivatives may be obtained by differentiating $y = \ln x$, yielding

$$\frac{dy}{dx} = \frac{1}{x}$$

The value of the derivative at the required values of x are then

x	1	2	3	4	5
dy/dx	1	0.5	0.333333	0.25	0.2

We see that our method gave values of the derivative that were in agreement with the exact figures to within six significant figures with delta equal to 0.001 or smaller and that even the derivatives computed with a delta of 0.1 are accurate enough for many purposes.   ////

11-3 POLYNOMIAL INTERPOLATION

In Sec. 8-4 we saw that we could write an nth degree polynomial

$$y = C_0 + C_1 x + C_2 x^2 + C_3 x^3 + C_4 x^4 + \cdots + C_n x^n$$

to pass through any $n + 1$ points. We first write such an equation to pass through several points of our data in the region at which we wish to find the derivative. We then take the derivative of this interpolating polynomial. It will be approximately equal to the derivative of the actual function. Successive derivatives of the interpolating polynomial are easily found,

$$y' = C_1 + 2C_2 x + 3C_3 x^2 + 4C_4 x^3 + \cdots + nC_n x^{n-1}$$
$$y'' = 2C_2 + 6C_3 x + 12C_4 x^2 + \cdots + n(n - 1)C_n x^{n-2}$$
$$\cdots\cdots\cdots\cdots\cdots\cdots\cdots\cdots\cdots\cdots\cdots\cdots\cdots\cdots\cdots\cdots\cdots$$

where the values of the constants $C_0, C_1, \ldots, C_n$ depend upon the degree of the interpolating polynomial and are defined in Chap. 8.

EXAMPLE 11-2   Find the first derivatives of the function described by the following table of data points, at values of x from 0.5 to 5 in steps of 0.5, using three-point interpolation.

x	y
0.5	-0.6930
1.0	0
1.5	0.40547
2.0	0.69315
2.5	0.91629
3.0	1.09861
3.5	1.25276
4.0	1.38629
4.5	1.50408
5.0	1.60944

These points were actually chosen to lie on the curve $y = \ln x$ of the previous example, so that we may check our results by differentiation.

SOLUTION   Lagrange's polynomial for three-point interpolation is

$$y = C_0 + C_1 x + C_2 x^2 \tag{8-4}$$

Taking the first derivative, we get

$$\frac{dy}{dx} = C_1 + 2C_2 x$$

where C_1 and C_2 are, for equally spaced data points,

$$C_1 = \frac{-y_0(x_1 + x_2) + 2y_1(x_0 + x_2) - y_2(x_0 + x_1)}{2h^2}$$

$$C_2 = \frac{y_0 - 2y_1 + y_2}{2h^2} \tag{8-5}$$

The interval h in this example is $\frac{1}{2}$, so that

$$2h^2 = 2(\tfrac{1}{4}) = \tfrac{1}{2}$$

Let us write a program with nested loops. We shall set x_0 initially equal to the first abscissa in the table (0.5), and while interpolating over the first three points in the table we shall find the derivative at each of these points. Then we shall increase x_0 by 0.5 and interpolate over the second, third, and fourth points in the table, finding the derivatives at each of these points. We repeat these steps until we have interpolated over the last three points in the table. Thus, we shall obtain the derivative for most of the points three separate times: when the point is at the beginning of the interpolation range, when it is in the middle, and when it is at the end. It will be interesting to compare the slopes obtained in the different positions and to compare them with the exact values.   The flowchart for this computation is given in Fig. 11-4, and the results are as follows:

Abscissa	Derivative	Percent error
0.5	1.67353	16.3
1	1.09847	9.8
1.5	0.52341	21.5
1	0.92873	7.1
1.5	0.69315	4.0
2	0.45757	8.5
1.5	0.63990	4.0
2	0.51082	2.2
2.5	0.38174	4.6
2	0.48710	2.6
2.5	0.40546	1.4
3	0.32382	2.9
2.5	0.39281	1.8
3	0.33647	0.94
3.5	0.28013	2.0
3	0.32892	1.3
3.5	0.28768	0.6
4	0.24644	1.4
3.5	0.28280	1.0
4	0.25132	0.53
4.5	0.21984	1.1
4	0.24801	0.80
4.5	0.22315	0.41
5	0.19829	0.85

The figures in the "Percent error" column are the differences between the computed values and the exact values (from $dy/dx = 1/x$) divided by the exact values, and multiplied by 100.   Notice that the errors are fairly large and also that the error is usually lowest when the derivative is taken at the midpoint of the interpolation range.   ////

The large errors possible by this method are due to the fact that even though the interpolation polynomial passes exactly through the given points, its *slope* may be considerably different from the true slope at those points.   The situation can become much worse if the data contain some experimental errors as in Fig. 11-5. Line AB is drawn among the 10 data points to show the general trend of the data. If we now interpolate over any three points, such as P_1, P_2, and P_3 with the second-degree curve CD, it is clear from the diagram that the slopes obtained from CD will be much different from those from AB.

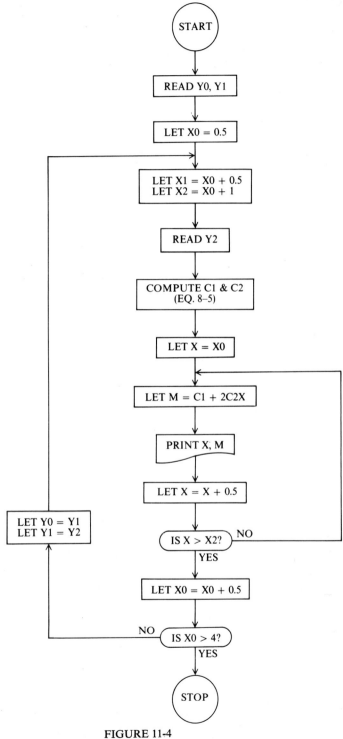

FIGURE 11-4
Flowchart for Example 11-2.

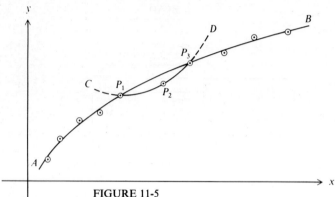

FIGURE 11-5
A possible source of error when using polynomial interpolation to find derivatives.

EXERCISE 11-1

1 Write a program to find first derivatives "by definition." Compute and print the first derivatives of the following functions from -5 to 5 in steps of 1, and plot the results. check your answers by manually differentiating the functions.

(a) $y = x^3 - 2x^2 + 4x + 3$ $\hspace{3cm}$ *Check* $y'(0) = 4$

(b) $y = \dfrac{3x^2}{8x^3 - 6}$ $\hspace{3cm}$ *Check* $y'(0) = 0$

(c) $y = 3 \sin 2x - 2 \cos 4x$ $\hspace{2.5cm}$ *Check* $y'(0) = 6$

(d) $y = 4 \sec 2x \tan 2x$ $\hspace{3cm}$ *Check* $y'(0) = 8$

(e) $y = 2x^3 - 3e^{2x}$ $\hspace{3.5cm}$ *Check* $y'(0) = -6$

2 Given the following table of data, determine the first derivative at each given data point, by polynomial interpolation. Plot your results.

x	0	1	2	3	4	5	6
y	0.79	2.91	4.17	4.91	5.09	4.61	2.02

Check $y'(2) = 1$

3 Write a program to find the maximum and minimum points between any two values of x, of any function. Do it by stepping along the x axis until a point is found where the first derivative changes sign, backing up one step and reducing the step size by a factor of 10, stepping forward again until the first derivative again changes sign, and repeating this

procedure until the maximum or minimum point is found to the desired degree of accuracy. Use your program to find the maximum and minimum points of the following functions:

(a) $y = -2x^3 + 15x^2 - 24x + 100$ *Ans.*   Max at (4,116), min at (1,89)

(b) $y = 2x^3 - 3x^2 - 12x$ *Ans.*   Max at $(-1,7)$, min at $(2,-20)$

PROJECTS

1   Given the cam of Project 1, Chap. 8, rotating at 1 r/s, compute the radial velocity and radial acceleration of a cam follower at the rim of the cam, in 20° intervals of cam rotation. Print the results in a four-column table giving cam displacement (degrees), time (seconds), follower radial velocity (inches per second), and follower radial acceleration (inches per second squared).   Plot your results.

Check   Velocity = 7.88 in./sec and acceleration = 28.8 in./sec² when the displacement = 180°

2   A rectangular swimming pool having an area of 600 ft² is to be surrounded by an apron 4 ft wide along two parallel sides, and 3 ft wide along the other two sides.   The apron is to be made of expensive tiles.   Write a program to find the pool dimensions that will require the least amount of tile for the apron.   *Ans.*   21.2 × 28.3 ft

12

INTEGRATION

The solution of many technical problems depends upon being able to set up and evaluate *definite integrals,* of the form

$$\int_a^b f(x)\, dx$$

Typical of these problems are those involving the finding of areas, lengths of arcs, moments of inertia, centroids, etc.   If we know the function $f(x)$ and can integrate it, we may use the methods of integral calculus to reach a solution.   But, unfortunately, many functions cannot be integrated.   Still worse, we may not have an equation at all, with the relationship given instead in the form of a table of data.

Because of the frequent occurrence of this type of problem, many approximate methods have been devised to evaluate definite integrals.   They all depend upon the fact that the definite integral, as given in the above equation, can be interpreted as the area bounded by the curve $y = f(x)$, the x axis, and the lines $x = a$ and $x = b$.   Thus the problem of evaluating a definite integral boils down to one of *finding an area,* even though the original physical problem for which we wrote the definite integral may have had nothing whatever to do with areas.

There are several graphical and mechanical methods for finding areas.   One method is to plot the figure on graph paper having small divisions and to find the area by counting the squares within the figure and applying appropriate scale factors. Land areas on aerial photographs are often measured by placing a transparent dot grid over the photograph and then counting the dots within the boundaries.   Another method is to paste the map to heavy sheet stock, cut out the required area, and weigh it on a sensitive balance.   Mechanical devices for measuring areas are called *planimeters*, and many designs are in existence.   A photoelectric planimeter determines areas by measuring the amount of light blocked by a cutout of the required figure.

In addition to these nonmathematical methods there are several other techniques suitable for programming on a computer.   These are methods of *numerical integration* (sometimes called *quadrature*) and are described in detail in the following sections.

12-1   RECTANGULAR, AVERAGE ORDINATE, MIDPOINT, AND TRAPEZOID RULES

Let us suppose we want to find the area of the figure bounded by the x axis, the lines $x = a$ and $x = b$, and some curve defined by a set of data points $P_0, P_1, P_2, \ldots, P_n$ (which may or may not be equally spaced) as in Fig. 12-1.   We may obtain an approximation to this area by use of the *rectangular rule*.   We divide our total area into a number of rectangular *panels*, as in Fig. 12-2, and then sum the areas of the rectangular panels:

$$A = y_0(x_1 - x_0) + y_1(x_2 - x_1) + \cdots + y_{n-1}(x_n - x_{n-1}) \tag{12-1}$$

where $(x_0, y_0), (x_1, y_1), \ldots, (x_n, y_n)$ are the coordinates of our data points P_0, P_1, $P_2, \ldots, P_n$.

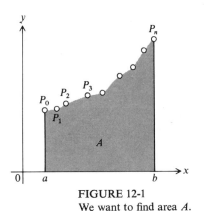

FIGURE 12-1
We want to find area A.

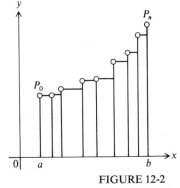

FIGURE 12-2
Rectangular rule.

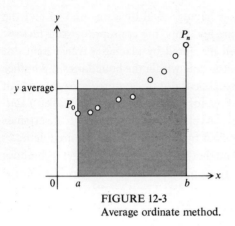

FIGURE 12-3
Average ordinate method.

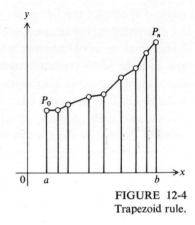

FIGURE 12-4
Trapezoid rule.

We can see from the figure that this will not be a good approximation to the total area unless a large number of data points are available. This method is rather crude and is presented here only for the sake of completeness.

The *average ordinate method* for approximating the area under a curve consists of finding the area of a rectangle whose width is $b - a$ and whose height is the average of the ordinates of all the data points, as in Fig. 12-3. The average ordinate is

$$y_{\text{avg}} = \frac{y_0 + y_1 + y_2 + \cdots + y_n}{n + 1} \qquad (12\text{-}2)$$

and the area is then

$$A = (b - a)y_{\text{avg}} \qquad (12\text{-}3)$$

If the data points are not well distributed over the total range but are clustered toward one end, this method will not yield good results.

Using the *trapezoid rule*, we approximate the curve between data points by straight-line segments, as in Fig. 12-4, and compute the area of each trapezoidal panel thus formed. The total area is then the sum of the areas of each of the trapezoids.

$$A = \tfrac{1}{2}[(x_1 - x_0)(y_1 + y_0) + (x_2 - x_1)(y_2 + y_1) + \cdots + (x_n - x_{n-1})(y_n + y_{n-1})] \qquad (12\text{-}4)$$

For equal intervals h, this equation reduces to

$$A = h[\tfrac{1}{2}(y_0 + y_n) + y_1 + y_2 + y_3 + \cdots + y_{n-1}] \qquad (12\text{-}5)$$

If out data points are equally spaced in x, or if we know the equation of the curve so that we may divide the interval into an even number of equally spaced panels, we may use the *midpoint method*. The area of each *double* panel is approxi-

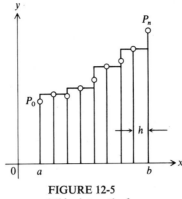

FIGURE 12-5
Midpoint method.

mated as a rectangle whose width is $2h$ and whose height is the ordinate at the midpoint of the double panel, as in Fig. 12-5. The total area is then

$$A = 2h(y_1 + y_3 + y_5 + \cdots + y_{n-1}) \tag{12-6}$$

All the above equations are easily programmed. The rectangular rule is the least accurate, is not easier to program than the others, and should be avoided. For unevenly spaced data points the trapezoid rule is preferred. When the equation of the curve is known so that we may take the intervals as small as we please, we can take smaller and smaller intervals in successive computations of the area. We then stop the run when the change in the computed area is less than some desired amount.

EXAMPLE 12-1 Suppose we ran a laboratory experiment and obtained the 19 unequally spaced data points given in the following table:

x	y
0	10.0
0.54	7.05
0.73	6.15
1.42	3.49
1.80	2.44
2.15	1.72
2.71	1.08
3.40	1.16
3.72	1.52
4.04	2.08
4.85	4.42
5.41	6.81
6.25	11.7
6.83	15.6
7.05	17.4
7.43	20.6
7.81	24.1
8.52	31.5
9.00	37.0

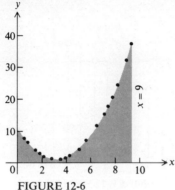

FIGURE 12-6
Plot of data in Example 12-1.

and we wish to find the area bounded by the curve defined by these points, the x and y axes, and the line $x = 9$, as shown in Fig. 12-6.

By the rectangular rule [Eq. (12-1)] we get

$$A = 10.0(0.54 - 0) + 7.05(0.73 - 0.54)$$
$$+ 6.15(1.42 - 0.73) + \cdots + 31.5(9 - 8.52) = 82.5$$

By the average ordinate method [Eq. (12-3)] we get

$$A = \frac{(9 - 0)(10 + 7.05 + 6.15 + \cdots + 37)}{19} = 97.5$$

And by the trapezoid rule [Eq. (12-4)] we get

$$A = \tfrac{1}{2}[(0.54 - 0)(7.05 + 10) + (0.73 - 0.54)(6.15 + 7.05)$$
$$+ (1.42 - 0.73)(3.49 + 6.15) + \cdots + (9 - 8.52)(37 + 31.5)]$$
$$= 90.6$$

Now, the data points in this example are not actual laboratory data but are points on the curve

$$y = x^2 - 6x + 10$$

They were so selected for the purpose of this example so that we may check our approximate solutions by integration. Integrating the above equation, we get

$$A = \int_0^9 (x^2 - 6x + 10)\, dx = \frac{x^3}{3} - \frac{6x^2}{2} + 10x \Big]_0^9 = 90$$

which is the exact value of the desired area. We see that the trapezoid rule gave the closest approximation, and this is generally the case.   ////

12-2   SIMPSON'S RULE

With the trapezoid rule, we approximated our curve between successive data points
with a straight line, in much the same manner as for linear interpolation.   In the
chapter on interpolation, we saw we could get a better approximation to the true curve
by connecting our data points with a higher-order equation, and the Lagrange interpolating formula was used for this purpose.   We shall use that formula again to
connect three of our data points with a second-degree equation and shall then find the
area of the two panels under that curve.   Once that procedure is established, it can be
repeated for each successive set of panels until the entire area has been computed.

Let us take three data points with *evenly spaced* abscissas, (x_0, y_0), (x_1, y_1),
(x_2, y_2).   Since a shift in the x axis will not affect the area under the curve between
x_0 and x_2, let us take the y axis through the middle point so that

$$x_0 = -h$$
$$x_1 = 0$$
$$x_2 = h$$

where h is the interval between successive values of x (Fig. 12-7).   The interpolating
polynomial for three equally spaced points is

$$y = C_0 + C_1 x + C_2 x^2 \tag{8-4}$$

where

$$C_0 = \frac{-y_1(-h)h}{h^2} = y_1$$

$$C_2 = \frac{y_0 - 2y_1 + y_2}{2h^2} \tag{8-5}$$

(C_1 is not needed.)
Integrating Eq. (8-4) between the limits $-h$ to h, we get

$$A = C_0 x + \frac{C_1 x^2}{2} + \frac{C_2 x^3}{3} \Bigg]_{-h}^{h}$$

$$= C_0(h + h) + \frac{C_1(h^2 - h^2)}{2} + \frac{C_2(h^3 + h^3)}{3}$$

$$= 2hy_1 + \frac{2h^3(y_0 - 2y_1 + y_2)}{6h^2} \tag{12-7}$$

$$A = \frac{h}{3}(y_0 + 4y_1 + y_2)$$

which is the well-known *Simpson's rule*.   Notice that the area does not depend upon

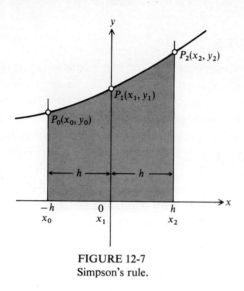

FIGURE 12-7
Simpson's rule.

the values of x but only upon the ordinates of the three data points and the common interval h.

If the data are in the form of an equation we should divide the total interval into an *even* number of panels, since, by this method, we integrate over two panels at once. If the data are in the form of a table with an odd number of intervals, one panel will be left over. If we assume the last panel to be a trapezoid and so compute its area, there will be little loss in accuracy of the entire integral.

As with the previous methods, if our data are in the form of an equation we have the further option of choosing our interval small enough so as to get any degree of accuracy. For any given number of intervals, Simpson's rule will give greater accuracy than any of the methods previously mentioned.

Application of Simpson's rule to unevenly spaced data becomes quite complicated, and it is recommended that the trapezoid rule be used instead.

EXAMPLE 12-2   Let us find the same area that we computed in Example 12-1, but this time let us assume 19 *evenly* spaced data points so that we may use Simpson's rule and the midpoint method, as well as the other three methods. Suppose our experiment yielded the following table of data:

x	y
0	10
0.5	7.25
1	5
1.5	3.25
2	2
2.5	1.25
3	1
3.5	1.25
4	2
4.5	3.25
5	5
5.5	7.25
6	10
6.5	13.25
7	17
7.5	21.25
8	26
8.5	31.25
9	37

By the rectangular rule [Eq. (12-1)] we get

$$A = (10 + 7.25 + 5.00 + 3.25 + \cdots + 31.25)0.5 = 83.6$$

By the average ordinate method [Eqs. (12-2) and (12-3)] we get

$$A = \frac{(10 + 7.25 + 5.00 + \cdots + 37.00)(9 - 0)}{19} = 96.8$$

By the trapezoid method [Eq. (12-5)] we get

$$A = 0.5[\tfrac{1}{2}(10 + 37) + 7.25 + 5.00 + \cdots + 31.25] = 90.4$$

By the midpoint method [Eq. (12-6)] we get

$$A = 2(0.5)(7.25 + 3.25 + 1.25 + \cdots + 31.25) = 89.3$$

By Simpson's rule [Eq. (12-7)] we get

$$A = \frac{0.5}{3}\{[10 + 4(7.25) + 5] + [5 + 4(3.25) + 2]$$
$$+ [2 + 4(1.25) + 1]\cdots + [26 + 4(31.25) + 37]\} = 90$$

We see that Simpson's rule gives the exact value of the area in this example. This is what we would expect since the equation of the given curve and the approximating equation are both of second degree, and we get a perfect fit. This would not be the case if our equation were other than second degree.   ////

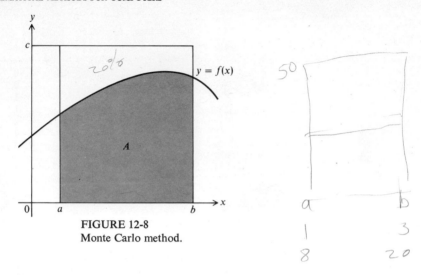

FIGURE 12-8
Monte Carlo method.

12-3 MONTE CARLO METHOD

Techniques using sets of random numbers can often provide interesting solutions to a variety of problems. Such techniques are generally referred to as the *Monte Carlo method* and are quite impractical without a high-speed computer. We shall use the method to find the definite integral of the function $y = f(x)$ between the limits a and b.

Let us enclose the desired area entirely within a rectangle whose height is c and width is $b - a$, having an area $A_r = (b - a)c$ as in Fig. 12-8. Let us now place points at random within this rectangle by generating random-number pairs (x_r, y_r), making sure that x_r falls within the range a to b and that y_r is less than or equal to c. Then we note whether our randomly located point is within the area A (a hit) or not (a miss) by comparing the random ordinate y_r with the computed ordinate $y = f(x_r)$. We generate many such randomly placed points, keeping a tally of our hits and of the total number of trials, and our area will be

$$A = \frac{total\ hits}{total\ trials} A_r$$

We may continue to add points until there is no further improvement, within our required accuracy, in the computed area. Note that, in order to use this method, the equations of the bounding curves must be known.

EXAMPLE 12-3 Let us take our previous example and find the area by the Monte Carlo method. We shall enclose this area within a rectangle of width 9 and height 37, whose area is then $9(37) = 333$. We shall place a point at random within this rectangle and see if it lies within the desired area by noting whether the ordinate of this random point is greater than $x^2 - 6x + 10$ at the same abscissa. We shall do

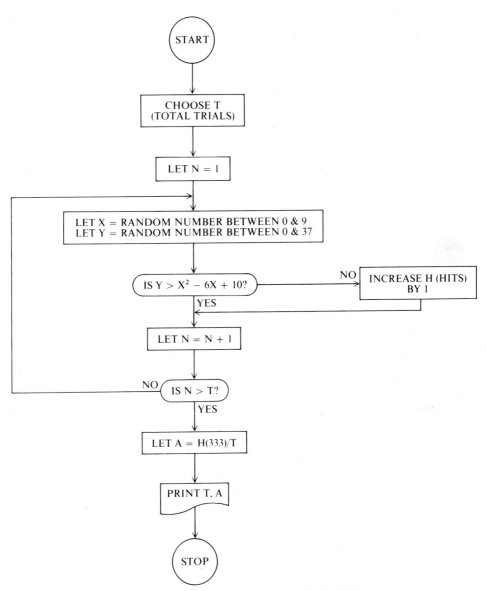

FIGURE 12-9
Flowchart for the Monte Carlo method.

this T times, for different values of T to see how the accuracy varies as we change the number of trials. The flowchart for the computation is given in Fig. 12-9. The computation was repeated for 10, 100, 1000, 10,000, and 50,000 trials, and the results are as follows:

Trials	Area
10	133.2
100	119.9
1000	85.9
10,000	90.3
50,000	89.8

////

EXERCISE 12-1

1 Write a program to compute the area under a given curve between given limits, using any of the methods in this chapter. Try your program on the following integrals.

(a) $\displaystyle\int_{1}^{8} (4x^{1/3} + 3)\, dx$ *Ans.* 66

(b) $\displaystyle\int_{4}^{9} \frac{1}{\sqrt{x}}\, dx$ *Ans.* 2

(c) $\displaystyle\int_{0}^{1} x\sqrt{1 - x^2}\, dx$ *Ans.* $\frac{1}{3}$

(d) $\displaystyle\int_{1}^{5} \frac{x + 6}{\sqrt{x + 4}}\, dx$ *Ans.* $\frac{50}{3}$

2 Given the data points of Exercise 11-1, Prob. 2, find the area under the curve by Simpson's rule.

PROJECTS

1 Given the light source and filter of Example 8-5, compute the total energy passing through the filter, in watts, by integrating the product curve given in Table 8-4. *Ans.* 8658 W
2 A Fourier series may be used to express any periodic function as the sum of an infinite number of sinusoidal waves of different frequencies. Such a series may be written

$$y = \frac{a_0}{2} + a_1 \cos x + a_2 \cos 2x + a_3 \cos 3x + \cdots + b_1 \sin x + b_2 \sin 2x + b_3 \sin 3x + \cdots$$

where the coefficients are found from the expressions

$$a_n = \frac{1}{\pi} \int_{0}^{2\pi} y \cos nx\, dx$$

$$b_n = \frac{1}{\pi} \int_{0}^{2\pi} y \sin nx\, dx$$

These two integrals may be evaluated for several values of n if y is known as a function of x, and if the products $y \cos nx$ and $y \sin nx$ are integrable. If an equation for y is not available, the integration must be performed by methods such as those given in this chapter.

A voltage wave is obtained from an oscillogram and its ordinates measured at equal intervals of the abscissa, resulting in the following table.

	x	y
	0	0
Note: Data is given for the	$\pi/16$	.99
first half-cycle only. For the	$\pi/8$	1.91
second half-cycle (from π to	$3\pi/16$	2.74
2π) the ordinates have the	$\pi/4$	3.18
same magnitude but are	$5\pi/16$	3.35
negative.	$3\pi/8$	3.31
	$7\pi/16$	3.15
	$\pi/2$	2.92
	$9\pi/16$	2.76
	$5\pi/8$	2.64
	$11\pi/16$	2.58
	$3\pi/4$	2.50
	$13\pi/16$	2.34
	$7\pi/8$	2.03
	$15\pi/16$	1.21
	π	0

Write a program to compute the areas under the curves $y \cos nx$ and $y \sin nx$ for values of n from 0 to 10, and use these areas to compute the first 10 Fourier coefficients a_n and b_n.

Check $a_1 = 0.1557$, $b_1 = 3.4915$

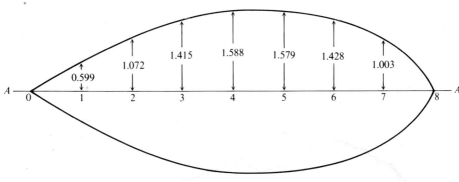

DIMENSIONS ARE IN INCHES
FIGURE 12-10
Cross section of a strut.

3   The streamlined strut whose cross section is shown in Fig. 12-10 is to be used in wind-tunnel tests.   It is necessary to know its cross-section area and its moment of inertia about the axis of symmetry *A-A*.   Determine both by means of Simpson's rule.

Ans.   Area $= 17.71$ in.^{2}.   Moment of Inertia $= 10.37$ in.^{4}

BIBLIOGRAPHY

CALINGAERT, PETER: "Principles of Computation," Addison-Wesley Publishing Company, Inc., Reading, Mass., 1965.

CONTE, S. D.: "Elementary Numerical Analysis, an Algorithmic Approach," McGraw-Hill Book Company, New York, 1965.

KEMENY, JOHN G., and KURTZ, THOMAS E.: "Basic Programming," 2d ed., John Wiley & Sons, Inc., New York, 1971.

LEDLEY, ROBERT S.: "Programming and Utilizing Digital Computers," McGraw-Hill Book Company, New York, 1962.

LIPKA, JOSEPH: "Graphical and Mechanical Computation," John Wiley & Sons, Inc., New York, 1918.

LYTEL, ALLAN: "Fundamentals of Computer Math," The Bobbs-Merrill Company, Inc., Indianapolis, 1964.

ORGANICK, ELLIOTT I.: "A Fortran IV Primer," Addison-Wesley Publishing Company, Inc., Reading, Mass., 1966.

PENNINGTON, RALPH H.: "Introductory Computer Methods and Numerical Analysis," The Macmillan Company, New York, 1970.

POLYA, GEORGE: "How To Solve It," 2d ed., Doubleday & Company., Inc., Garden City, N. Y., 1957.

SCARBOROUGH, JAMES B.: "Numerical Mathematical Analysis," 6th ed., The Johns Hopkins Press, Baltimore, 1966.

SPIEGEL, MURRAY R.: "Schaum's Outline of Theory and Problems of Statistics," McGraw-Hill Book Company, New York, 1961.

STARK, PETER A.: "Introduction to Numerical Methods," The Macmillan Company, New York, 1970.

INDEX